IIW Collection

Series editor

IIW International Institute of Welding,
ZI Paris Nord II, Villepinte, France

The IIW Collection of Books is authored by experts from the 59 countries participating in the work of the 23 Technical Working Units of the International Institute of Welding, recognized as the largest worldwide network for welding and allied joining technologies.

The IIW's Mission is to operate as the global body for the science and application of joining technology, providing a forum for networking and knowledge exchange among scientists, researchers and industry.

Published books, Best Practices, Recommendations or Guidelines are the outcome of collaborative work and technical discussions-they are truly validated by the IIW groups of experts in joining, cutting and surface treatment of metallic and non-metallic materials by such processes as welding, brazing, soldering, thermal cutting, thermal spraying, adhesive bonding and microjoining. IIW work also embraces allied fields including quality assurance, non-destructive testing, standardization, inspection, health and safety, education, training, qualification, design and fabrication.

More information about this series at http://www.springer.com/series/13906

Bastien Chapuis · Eric Sjerve
Editors

Sensors, Algorithms and Applications for Structural Health Monitoring

IIW Seminar on SHM, 2015

INTERNATIONAL INSTITUTE OF WELDING
A world of joining experience

Springer

Editors
Bastien Chapuis
NDE Department
CEA LIST
Gif-sur-Yvette
France

Eric Sjerve
IRISNDT
Edmonton
Canada

ISSN 2365-435X ISSN 2365-4368 (electronic)
IIW Collection
ISBN 978-3-319-69232-6 ISBN 978-3-319-69233-3 (eBook)
https://doi.org/10.1007/978-3-319-69233-3

Library of Congress Control Number: 2017957667

Printed on acid-free paper

This Springer imprint is published by Springer Nature
The registered company is Springer International Publishing AG
The registered company address is: Gewerbestrasse 11, 6330 Cham, Switzerland

Preface

Structural health monitoring (SHM) involves introducing a sensor, or an array of sensors, into an engineered structure to monitor it periodically for structural degradation due to its operating environment. The sensors provide nondestructive measurements to give information about critical structural properties, such as wall thickness measurements for corrosion detection, crack initiation or growth monitoring for welds susceptible to cracking, vibration monitoring of a structure that is susceptible to fatigue, and stress measurements for structures where either peak stresses or detailed stress measurements must be monitored. This data is then statistically analyzed to determine the current health of the structure, to estimate its remaining life, and to allow decisions to be made for follow-up maintenance.

SHM is a multidisciplinary field that is currently changing rapidly due to technological advances, and one that currently does not have well-established standardization. It is an important field for the IIW, as all welded or otherwise joined structures require monitoring to extend their lifetimes and to ensure continued safe operation.

Commissions V (Non-destructive Testing and Quality Assurance of Welded Products), XI (Pressure Vessels, Boilers and Pipelines), XIII (Fatigue of Welded Components and Structures), and XV (Design, Analysis and Fabrication of Welded Structures) held a joint seminar during 68th IIW annual assembly 2015 in Helsinki (Finland) on this important subject. Many countries and industries around the world are managing aging infrastructure, and examining ways of extending life within economic constraints and requirements for public and environmental safety. The attendance by over 60 representatives from 17 countries demonstrated the importance of this forum and the knowledge shared by attendees.

Papers were presented in the area of sensor development, including macro-fiber composite sensors for crack detection and optical fiber Bragg gratings for flaw detection. The use of welds itself in the structure as SHM sensors and simulation to determine the probability of detection for different SHM sensor configurations were also discussed. SHM methods based on vibration signal variations to detect small defects in composite components or to monitor large structures (hull of a container ship) were considered. There were also application-specific papers describing SHM applied to industrial components such as a nuclear boiler support spine, an industrial press, and corrosion monitoring of pipes.

This booklet gathers together several papers related to presentations made during this joint SHM seminar. It will allow for the attendees to go further on the subject than what has been presented during the seminar and for the other readers to discover recent developments in that discipline. Following the success of this first seminar, IIW will continue to have involvement in SHM through seminars and joint commission activities.

Gif-sur-Yvette, France Bastien Chapuis
Edmonton, Canada Eric Sjerve

Contents

Contributors

Frederic Cegla Department of Mechanical Engineering, Imperial College London, London, UK

Bastien Chapuis NDE Department, CEA LIST, Gif-Sur-Yvette, France

Daniel Chauveau Innovation Department, Institut de Soudure Industrie, Villepinte, France

Fethi Dahmene Institut de Soudure, Plateforme RDI CND, Yutz, France

Tom Druet NDE Department, CEA LIST, Gif-sur-Yvette, France

Alexander Dubov Energodiagnostika Co. Ltd, Moscow, Russia

Anatoly Dubov Energodiagnostika Co. Ltd, Moscow, Russia

Mahjoub El Mountassir Institut de Soudure, Plateforme RDI CND, Yutz, France

Manfred Jules CEA LIST, Gif-sur-Yvette, France

Adrian Kahl DNV GL, Hamburg, Germany

M. H. Kim Department of Naval Architecture and Ocean Engineering, Pusan National University, Busan, South Korea

Sergey Kolokolnikov Energodiagnostika Co. Ltd, Moscow, Russia

Guillaume Laffont CEA LIST, Gif-sur-Yvette, France

J. H. Lee Department of Naval Architecture and Ocean Engineering, Pusan National University, Busan, South Korea

J. M. Lee Department of Naval Architecture and Ocean Engineering, Pusan National University, Busan, South Korea

Emmanuel Moulin Université de Valenciennes et du Hainaut-Cambrésis, Valenciennes, France

Peter Mudge TWI Ltd., Cambridge, UK

D. J. Oh Department of Naval Architecture and Ocean Engineering, Pusan National University, Busan, South Korea

Hubertus von Selle DNV GL, Hamburg, Germany

Gaute Storhaug DNV GL, Hamburg, Germany

Slah Yaacoubi Institut de Soudure, Plateforme RDI CND, Yutz, France

Abstract

Structural health monitoring (SHM) involves introducing a sensor, or an array of sensors, into an engineered structure to monitor it periodically for structural degradation due to its operating environment. SHM is a multidisciplinary field that is currently changing rapidly due to technological advances, and one that currently does not have well-established standardization. It is an important field for the IIW, as all welded or otherwise joined structures require monitoring to extend their lifetimes and to ensure continued safe operation.

Commissions V (Non-destructive Testing and Quality Assurance of Welded Products), XI (Pressure Vessels, Boilers and Pipelines), XIII (Fatigue of Welded Components and Structures), and XV (Design, Analysis and Fabrication of Welded Structures) held a joint seminar during 68th IIW annual assembly 2015 in Helsinki (Finland) on this important subject.

This booklet gathers together several papers related to presentations made during this joint SHM seminar. It will allow for the attendees to go further on the subject than what has been presented during the seminar and for the other readers to discover recent developments in that discipline.

Keywords Nondestructive testing · SHM · Structural health monitoring · Sensors Service conditions

Introduction to Structural Health Monitoring

Bastien Chapuis

Abstract Structural Health Monitoring (SHM) consists of embedding sensors in a structure to periodically monitor its health. The main motivations for this are to increase safety, optimize, and anticipate maintenance operations, and to extend the working life of the structure. Longer term applications of SHM consist of reducing mechanical safety margins with subsequent weight savings. SHM emerges from the field of smart structures and encompasses various disciplines such as materials, fatigue, nondestructive testing, sensors, signal processing, communication systems, and decision-making processes. It is an emerging field, with a very active research community, and the initial industrial deployments are being demonstrated as effective. SHM is at the beginning of the structuring process, with standardization of the techniques and methodologies still being worked on. However, major developments are expected in the near future due to the current tremendous progress of embedded device technologies in daily use brought about by Internet of Things applications. In this short introductory paper, the current challenges of SHM are illustrated with some examples coming from different industrial applications.

Keywords NDT · SHM · Structural health monitoring

1 Introduction

Structural Health Monitoring (SHM) is an emerging field in engineering that gathers together several techniques such as structural dynamics, materials, signal processing, or microelectronics that can be used to give at every moment during the life of a structure a **diagnosis** of its health and a **prognosis** of its remaining life.

V-1667-15 (XV-1493-15)

B. Chapuis (✉)
NDE Department, CEA LIST, Gif-sur-Yvette, France
e-mail: bastien.chapuis@cea.fr

B. Chapuis and E. Sjerve (eds.), *Sensors, Algorithms and Applications for Structural Health Monitoring*, IIW Collection, https://doi.org/10.1007/978-3-319-69233-3_1

This is done through a full system that comprises sensors and actuators, communication systems, and data processing units, from which part of it (at least the sensors) are embedded in the structure itself.

In its first function, diagnosis, SHM is very similar to Non-Destructive Testing (NDT). Indeed, these two fields use common physical techniques of investigation, and the interaction between the two domains and communities are numerous, at least in order to compare the results and to validate the approach if necessary.

But SHM must go further than NDT. First, its "embedded" aspect induces sufficient constraints that it is generally impossible to simply miniaturize an existing NDT system and integrate it into the structure. The global monitoring strategy must generally be completely reconsidered, sometimes back to design of the structure. Second, the notion of "prognosis", i.e., estimating the residual life of the structure (or at least anticipating the flaw growth evolution until the next maintenance operation), is fully included into SHM. This is a very complex aspect that requires a deep knowledge of material damage mechanisms. Thus, SHM is not only a damage detection process but also includes monitoring of its history and of the environment to which it has been submitted. At least, an SHM system must be able to detect an event above a predefined threshold in order to trigger a further NDT inspection.

However, SHM systems will not eliminate completely "classical" NDT. First, it seems impossible to design a set of systems that can monitor the entirety of a structure as complex as a plane or a bridge. Second, the precision of the diagnosis obtained with an integrated system will often be insufficient. Because of this, SHM is more likely to be an alarm system to identify zones where further inspection is required in order to facilitate and predict maintenance operations. NDT and SHM are therefore complementary, and a full health management strategy must be set up in order to take into account the advantages and limitations of both methods.

The two main domains that are the origin of this discipline are aeronautics and civil engineering (especially for bridges). Nowadays, SHM is commonly applied in other disciplines such as wind turbines, oil and gas (pipes), and railways.

A good synthesis of the state of the art of the discipline is presented since 1997 every odd year in the *International Workshop on Structural Health Monitoring* (*IWSHM*) in Stanford (California) and every even year in the *European Workshop on Structural Health Monitoring* (*EWSHM*) in a different country (France-2002, Germany-2004, Spain-2006, Poland-2008, Italy-2010, Germany-2012, France-2014).

The success of these congresses is growing, which illustrates the interest given to the discipline (see Fig. 1).

The physical implementation of an SHM system is very specific to the application and the structure being monitored. We will present in the next part the major aspects of SHM: the objectives and the technical challenges.

Then, some examples of recently developed SHM systems will be briefly described. These few examples will be advantageously completed by the other papers that are included in this booklet.

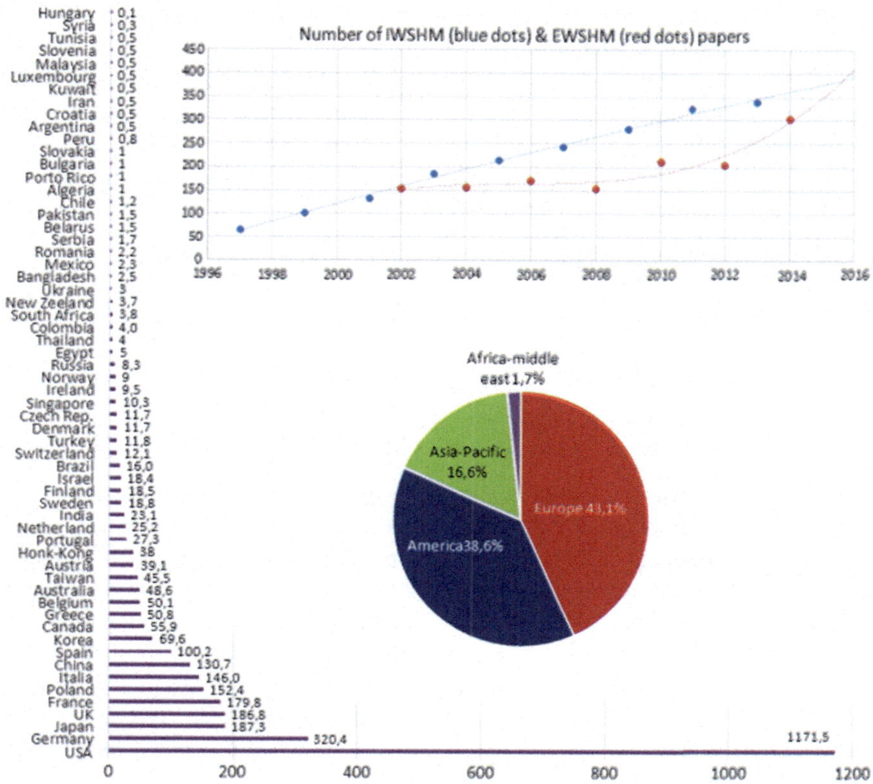

Fig. 1 SHM papers per origin (countries) and per event from IWSHM1999 to EWSHM2014 [6]

2 Objectives and Technical Challenges

2.1 Definitions and Objectives

A document published by SAE International in 2013, ARP6461 "Guidelines for Implementation of Structural Health Monitoring on Fixed Wing Aircraft", defines SHM as "The process of acquiring and analyzing data from onboard sensors to evaluate the health of a structure" [1].

More than the definition by itself, this guideline is very interesting since it is the first document written by the SHM community to standardize terminology and methodologies (for aerospace) after years of discussion. It was written by the SAE G-11 SHM Aerospace Industry Steering Committee, and it is recognized as a major step toward the transfer of SHM technologies from laboratories to real field environments.

The main objective of an SHM system is to ensure the health of the structure until the next maintenance operation. From the information that the system delivers, the user wants to optimize the use of the structure both for safety and in terms of increased operational availability.

Indeed, as part or all of the SHM system is embedded in the structure, the inspection might possibly be automated and repeated as often as possible. Provided that the information delivered are reliable, this should prevent catastrophic failures and unanticipated downtimes. The automation of the diagnosis should also limit the influence of human subjective interpretation of results that occur in some “classical” inspection operations.

Five levels of information can be given by the SHM system:

- Level 1: detection of damage,
- Level 2: location of the damage,
- Level 3: characterization (type) of the damage,
- Level 4: sizing (severity) of the damage, and
- Level 5: remaining life of the structure (prognosis).

SHM is often viewed as a way to simplify and/or anticipate maintenance operations. The system might be implemented in areas difficult to monitor due to inaccessibility problems. This is particularly true for some components of aircrafts that require significant dismantling work to provide accessibility, or in the case of offshore wind farms for which regular human presence is extremely difficult. However, the main advantage of SHM consists in anticipating downtimes of the structure, which offers large potential gains in terms of logistics.

SHM systems can also be used to increase the lifetime of a structure. As an example, this is the case in civil engineering where aging bridges in different countries all over the world are being equipped with monitoring systems to extend their working life.

Different strategies can be used to design an efficient SHM system, and they can be grouped into three categories as presented in Fig. 2.

The first one consists of monitoring the operational environment of the structure, which is called “operational monitoring” in ARP6461. In this case, parameters such as temperatures, loads, etc. are then used to build an indirect evaluation of structure usage. The main difficulty of this approach consists of the exhaustive list of monitored parameters that might contribute to structural degradation.

A second approach aims at directly monitoring the integrity of the material, i.e., looking for defects like in NDT. This requires very specialized sensors, which are generally close to the inspected area. This is called “damage monitoring” in ARP6461. The reliability of the inspection system must be carefully demonstrated (probability of detection and inspected area) to ensure the SHM system is providing accurate and useful data.

A third approach, applicable mainly for large structures such as bridges or wind turbines, consists of monitoring the shape of the structure to detect deformations outside tolerable limits that might indicate failure of a subcomponents or usage of the structure beyond the working limits it has been designed for.

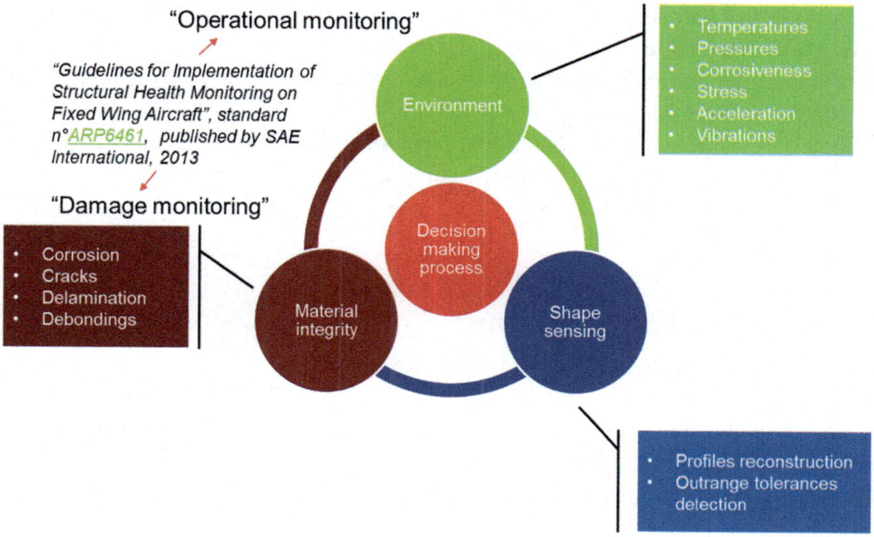

Fig. 2 Main categories of monitoring strategies

A single SHM system can comprise several types of sensors of different kinds, and whatever the strategies chosen, the data must be merged, processed, and evaluated through different advanced algorithms to ensure the structure's health.

2.2 *Technical Challenges*

Up to now, several technical challenges have limited the deployment of SHM systems in industry. The first one is the difficulty to design such complex SHM systems that require different technologies that are not always commercially available (sensors, data processing units, communications systems, etc.). All of these components must perform **reliably** (i.e., with good performance and acceptable false alarm rates) during at least the working life of the structure. **Durability** concerns must therefore often be solved since SHM systems might be deployed in harsh environments. **Integration** of the system in the structure must also not be a source of supplementary defects. This is especially the case when sensors like fiber optics are embedded in composite materials, and often additional testing might be required to demonstrate that. A big challenge in the design comes from the constraints that arise from each embedded system: limited **embedded processing capabilities**, limited access to energy or **energy storage,** and limited **data transmission bandwidth**. This generally requires the development of ad hoc solutions for data transmission protocols and energy management strategies.

In order to establish a **prognosis** of a structure's residual life, it is necessary to have good knowledge of material aging mechanisms and load distribution characteristics (time and intensity distribution probabilities) but this information might be difficult to determine.

Finally, even if ARP6461 is the first step, **standardization** of the technologies that are used in SHM and established **certification** methodologies are necessary before large adoption by industry.

3 Examples of SHM Systems

We will briefly present in this section three examples of SHM systems that are deployed in the field. This is far from being representative of the diversity of technologies that are used in SHM and more detailed examples will be illustrated in the other papers that have been presented in this seminar.

3.1 Wind Farms

An example of structures for which there has been a successful deployment of SHM system in industry is wind turbines. These structures are often deployed in very harsh environments (offshore, low temperatures, etc.) and they can be subjected to dramatic failures of blades (see Fig. 3) or tower. Maintenance operations can be

Fig. 3 Wind turbine blade failure [7]

Fig. 4 Blade inspection might require physical access to operators [10]

very difficult, meaning it is advantageous to replace human maintenance activities with automatic inspection systems (see Fig. 4).

Monitoring the vibrations of the wind turbine (either of the tower or of the blades) is a solution to anticipate some failures and to increase operational availability of the turbine.

For example, there is a risk of icing for wind turbines installed in Northern Europe or in mountain environments. Interruption of electricity production is required by authorities if ice thrown from blades poses a threat to people and animals. As an alternative to stopping production when temperature becomes too low, a solution proposed by the German company Wölfel consists of mounting on each blade at least one sensor unit consisting of a two-dimensional accelerometer and a temperature sensor [2].

Data processing is used to build an "ice indicator" that ensures, after compensation of temperature and operating conditions, a clear distinction between the normal vibrations of the blade and the structural response due to icing of the blade, as shown in Fig. 5. This allows regulating the operation of the wind turbine ensuring maximum energy production with the lowest risk of ice throw.

This successful example of SHM implementation is now deployed in several wind farms in Northern Europe.

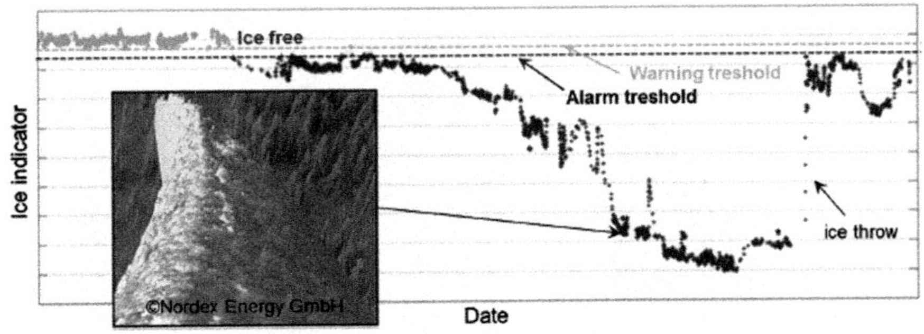

Fig. 5 Example of indication given by the system IDD.Blade® [2]

3.2 *Bridges*

Several bridges in the world have been equipped with instrumentation to monitor their structural health condition in terms of their environmental loads (wind, traffic, temperature, etc.) and response to evaluate and predict their current and future performance.

An example of a large-scale implementation of such a system is given by the so-called "Wind And Structural Health Monitoring System (WASHMS)" developed in Hong Kong in the past decade [3]. Five long-span cable-supported bridges have been instrumented, namely the Tsing Ma (suspension) Bridge, the Kap Shui Mun (cable-stayed) Bridge, the Ting Kau (cable-stayed) Bridge, the Western Corridor (cable-stayed) Bridge, and the Stonecutters (cable-stayed). A large number of sensors (up to 1700 per bridge) of different types have been installed at specific locations on the bridges (see Fig. 6). A modular architecture is able to acquire, process, and store the acquired data continuously, as shown in Fig. 7. With the acquired monitoring data, many investigations on condition assessment and damage detection of large-scale cable-supported bridges have been conducted [4].

3.3 *Aeronautics*

Along with civil engineering, aeronautics is one of the major fields for SHM investigations. However, at the present time, very few SHM systems have been implemented on aircraft and none is certified to replace conventional nondestructive inspections. One of the most advanced solutions is the so-called Comparative Vacuum Monitoring (CVM) system designed to detect small cracks (less than half a mm) at the surface of a component in known "hot spots" (for example near rivets). It consists of a thin self-adhesive rubber patch with interconnected galleries, to which air pressure is applied (see Fig. 8). Any propagating crack under the sensor breaches the galleries and the resulting change in pressure is monitored.

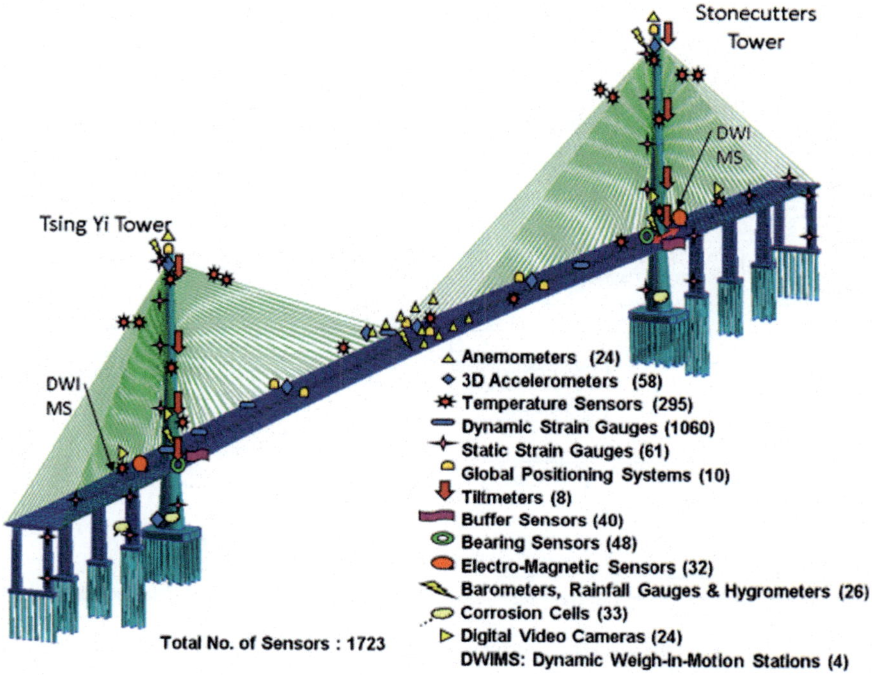

Fig. 6 Sensors installed on Stonecutters Bridge (Hong Kong) [4]

After years of research and trials with this technology, Delta/Boeing selected the 737 NG Wing Box fitting cracking problem as the application for a formal certification of CVM by the American Federal Aviation Administration (FAA) in 2015 [5].

4 Conclusions and Perspectives

Examples of SHM system deployment in industry are limited at the present time despite the promising benefits that can be expected in terms of safety and operational availability of the structure. This is mainly due to the very recent development of this discipline that integrates a variety of "classical" technologies to build a full and complex SHM system whose reliability must be proven before being installed on critical structures.

However, research in this field is very intense and several new applications should be demonstrated in the near future.

Major progress is expected by the use of **simulation** to design and demonstrate performance of SHM systems. Developments of new **sensors** (fiber optics, MEMS, etc.) or new inspection **methods** (guided waves) will increase the quantity and

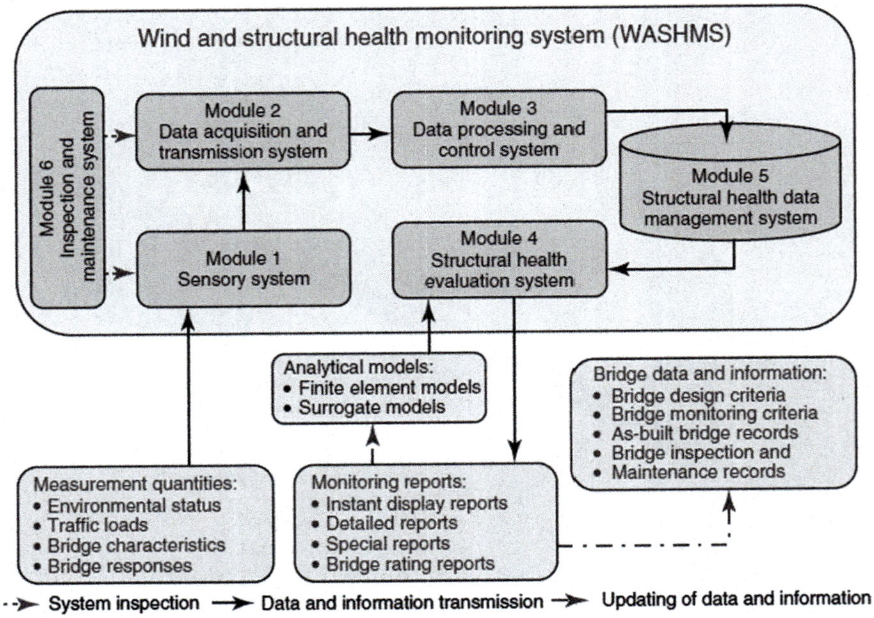

Fig. 7 Architecture of the wind and structural health monitoring system (WASHMS) deployed on Hong Kong bridges [3]

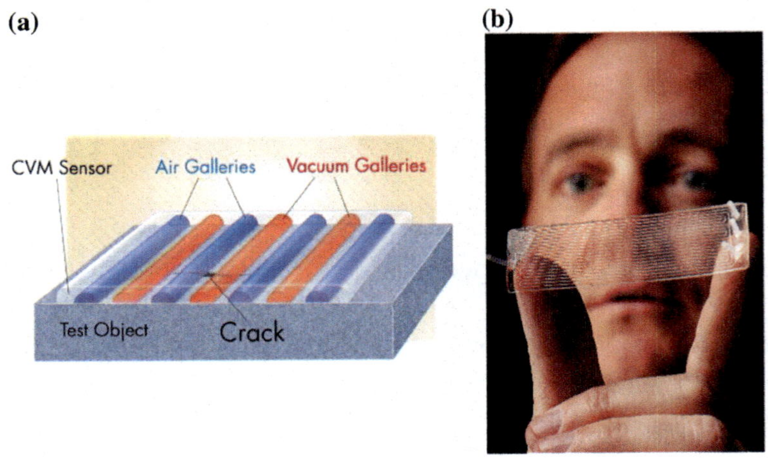

Fig. 8 **a** Schematics of a CVM sensor [8], **b** flexible CVM sensor [9]

quality of information that can be obtained from the structure. Combined with new data processing **algorithms** that take advantage of such a high volume of information (such as machine learning), this will lead to a more reliable ability to diagnose and predict a structure's health. New **architectures** of SHM systems, especially based on wireless networks, will also help to simplify on-site implementations.

Finally, SHM will also widely benefit of the current revolution of the Internet of Things (IoT), which is introducing sensors, memories, communication systems, and power resources in many objects of daily use. Making these technologies cheaper and standardized, it will accelerate the transfer of SHM from laboratory research into the industrial field.

References

1. SAE International (2013) Guidelines for Implementation of Structural Health Monitoring on Fixed Wing Aircraft, ARP6461
2. Kraemer P, Friedmann H, Ebert C (2015) Vibration-based ice detection of rotor blades in wind turbines—the industrial realization of an SHM-system. In: Proceedings of the 10th IWSHM, Stanford
3. Wong K-Y, Ni Y-Q (2009) Modular architecture of SHM system for cable-supported bridges, encyclopedia of structural health monitoring. Wiley, USA
4. Ni Y, Wong K (2012) Integrating bridge structural health monitoring and condition-based maintenance management. Civil Structural Health Monitoring Workshop, Berlin
5. Swindell P (2014) FAA structural health monitoring research program. Airlines for America NDT forum, Atlantic City
6. Ferdinand P (2014) The evolution of optical fiber sensors technologies during the 35 last years and their applications in structure health monitoring. In: Proceedings of the 7th EWSHM, Nantes
7. Health and Safety Executive (2013) Study and development of a methodology for the estimation of the risk and harm to persons from wind turbines, RR968
8. http://www.smsystems.com.au/content/products/to_how-it-works.asp
9. https://share.sandia.gov/news/resources/releases/2007/aircraft.html
10. http://www.offshorewind.biz/2014/09/03/rtn-unveils-turbine-blade-maintenance-courses/

Ultrasonic Monitoring of Corrosion with Permanently Installed Sensors (PIMS)

Frederic Cegla

Abstract Ultrasonic wall thickness monitoring using permanently installed sensors has become a tool to monitor pipe wall thicknesses online and during plant operation. This talk aims to give a short introduction to the benefits and drivers of the technology as well as the technical challenges that had to be overcome during the development of the waveguide sensor that is Permasense Ltd.'s flagship product. The presentation will present the underlying measurement principle and then give some more detail on key parameters that influence the measurement, such as temperature and surface morphology. It will also be shown how these effects can be managed. Some example data of application used in industry will be presented and the paper will end with a look into the near future and describe technology that will become available soon.

Keywords Ultrasonics · Corrosion · Structural health monitoring SHM

1 Introduction

Ultrasonic wall thickness measurements or ultrasonic thickness gauging is one of the most commonly employed tools to check that corrosion or erosion has not degraded metal work in industrial plant. The methods are well documented and there are industrial standards that describe how to carry out the measurements [1, 2]. Traditionally, all measurements were carried out manually. This means that an operator goes out to the location of the plant where the measurements are required; he gains access to the component (removing insulation, building scaffolding, or

---

V-1669-15.

F. Cegla (✉)
Imperial College London, London, UK
e-mail: f.cegla@imperial.ac.uk

B. Chapuis and E. Sjerve (eds.), *Sensors, Algorithms and Applications for Structural Health Monitoring*, IIW Collection, https://doi.org/10.1007/978-3-319-69233-3_2

rope access can be required) and then carries out the actual measurement. Because of the restricted access to some components, this procedure is not carried out frequently. Furthermore, because of coupling and positioning errors on repeat manual measurements, there is a rather large uncertainty on these measurements. Therefore, over the last decade, permanently installable, wireless thickness gauging sensors have been developed. Once installed, these sensors remain fixed in the same location and send back data at regular pre-configurable intervals. Several systems are on the market. The information relayed here will be kept as general as possible; however, the author has been deeply involved in the creation of the Permasense Ltd. wireless ultrasonic monitoring system, and therefore some information that is displayed might be biased toward this particular system [3].

It is of utmost importance that permanently installed sensors are as rugged as possible and therefore their coupling and attachment mechanisms are key. Furthermore, it needs to be ensured that the sensor can withstand the operating conditions (e.g., temperature) at the location where the transducer is to be installed. In the Permasense system, both these problems were solved by means of the use of a robust waveguide that isolates the fragile transducers and electronics from the measurement error. The waveguide is thin and slender so that it can isolate large temperature differences over short distances. The transducer is attached to the sample under test by means of two welded studs which apply load onto the contact patches and enable dry coupling of the waveguide to the pipe surface. This has proven to be much more reliable than bonding of the transducer.

This paper is organized as follows: Sect. 2 describes the key features of the fully wireless ultrasonic corrosion monitoring system as it is deployed in an industrial plant and shows some example long-term wall thickness trends. Section 3 is concerned with the effects of temperature on the monitored wall thickness and Sect. 4 is concerned with the effects of changes in surface morphology. Finally, conclusions are drawn from the presented information.

2 A Complete Industrial Ultrasonic Monitoring System

2.1 Overview of Components

Figure 1 shows an overview of the Permasense Ltd. wireless ultrasonic corrosion monitoring system. Sensors are installed at monitoring locations in the industrial plant, and they communicate wirelessly with each other (forming a mesh network) and relay data to a gateway that is connected to the general business IT network of the plant. Via the network the gateway communicates with a database server on which all the information is stored and users can visualize the data in the database from any office PC that has network access.

Fig. 1 Overview and pictures of the Permasense Ltd. wireless ultrasonic corrosion monitoring system

2.2 *A Typical Thickness Measurement*

Figure 2 shows the geometry of the waveguide sensor on a metal sample of thickness T. The measurement is a pitch-catch measurement from the transmitting transducer Tx to the receiving transducer Rx. The ultrasonic signal travels via different wave paths resulting in the arrival of distinct wave packets as can be seen in the ultrasonic signal of Fig. 2b. The arrival time difference between echoes is directly linked to the component thickness via the ultrasonic shear velocity in the material that is being tested. To estimate the component thickness, the arrival time difference is established and turned into a distance by multiplication with the wave velocity. The final result is a thickness value.

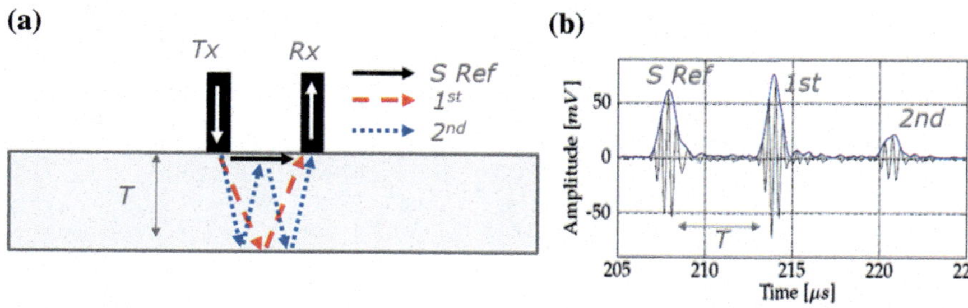

Fig. 2 **a** Illustration of ultrasonic signal path and **b** typical ultrasonic signal that is received by a sensor

2.3 Example Wall Thickness Measurement Trends

Figure 3 shows thicknesses that were monitored over the period of about 3.5 years in a plant component where the wall thickness remained constant. The plot demonstrates the very good repeatability and measurement frequency that can be obtained with a permanently installed monitoring system.

Figure 4 shows the wall thickness trend measured on another component in a plant. Here, it is clearly visible that wall loss is taking place. The wall loss rate is roughly 1 mm/year; however, there is also a period where no wall loss/corrosion is taking place. The measurement frequency and precision of the automated measurements make it possible to gather information on wall thickness changes that would not be obtained with conventional manual ultrasonic measurements. Conventional measurements would have much higher error bars (i.e., measurement uncertainty) and would record data much less frequently. This is indicated by the box plot on the right of Fig. 4 which illustrates ±1 standard deviation of the measurement uncertainty (indicated by the extent of the box; ±2 standard deviations for the whiskers) that has been quoted for manual UT measurements [4]. The additional information can be used to take decisions about corrosion mitigation strategies, the effect of operating conditions on the plant, or if component retirement should be considered in the near future.

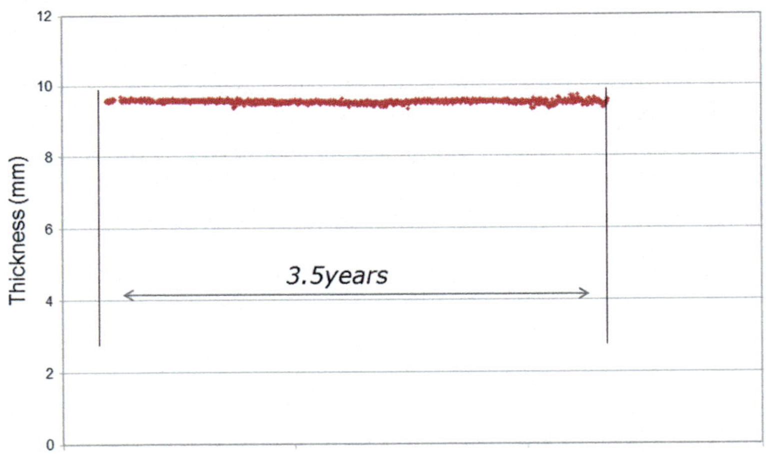

Fig. 3 Constant wall thickness monitored over the period of ~3.5 years

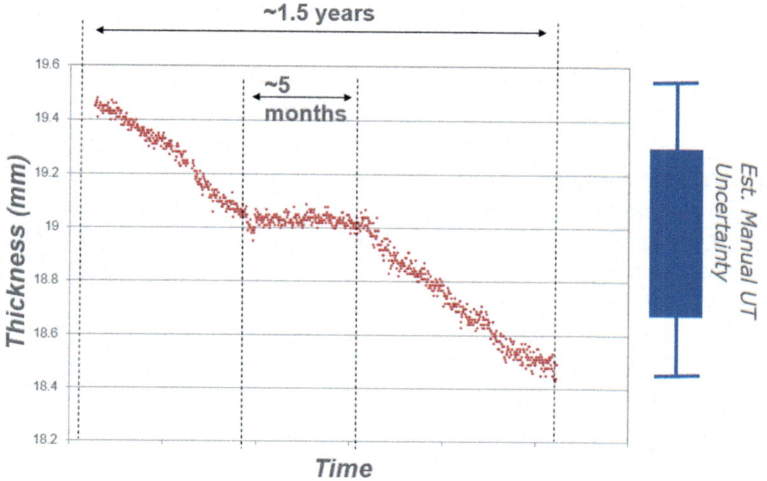

Fig. 4 Wall thickness trend showing wall loss due to corrosion and intermittent period of no corrosion, also indicated on the right is the uncertainty that is expected to result from manual ultrasonic measurements

3 Effects of Temperature on the Ultrasonically Monitored Wall Thickness

The ultrasonic wave velocity in steels is a function of temperature. This can introduce changes into wall thickness measurements, if these measurements are not made at the same temperature. Figure 5 shows the velocity temperature behavior of

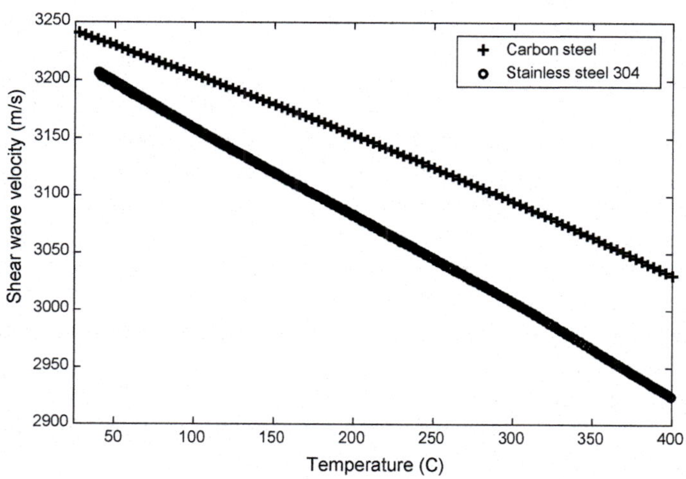

Fig. 5 Shear wave velocity as a function of temperature for carbon steel and stainless steel 304

a 10-mm-thick carbon steel plate and that of a 10-mm-thick stainless steel 304 plate that was measured in the laboratory. While overall the velocities are within 1% and both steels behave similarly, there are subtle differences in slope which can cause errors in the temperature compensation if large temperature swings are encountered or if a very good measurement precision is required. Based on the carbon steel measurements the ultrasonic velocity reduced as 0.56 m/s per °C or 0.017% per °C. This would cause a thickness error of ~0.17% or 17 μm in 10 mm wall thickness for a 10 °C temperature change. For the stainless steel, the results are slightly different and the velocity reduced as 0.78 m/s per °C or 0.024% per °C. Here, a thickness error of 24 μm in 10 mm wall thickness for a 10 °C temperature change would result. If the wrong calibration curve for carbon or stainless steel were used, an error of ~7 μm in 10 mm wall thickness for a 10 °C temperature change could result. Therefore, for very precise measurements, both temperature compensation and choice of the correct calibration data becomes important.

Gajdacsi [5] also analyzed the performance of temperature compensation of ultrasonic signals with a thermocouple that is attached to the outside of a fluid carrying pipe wall if the pipe wall is heated by the internal fluid. He showed that under steady temperature conditions there is little error in the temperature compensation when data from the external thermocouple is used. Under these conditions, sub-μm changes in wall thickness can be tracked [6]. However, if there is unsteady heating or cooling of the pipewall, then larger errors can result due to the nonuniform temperature distribution within the pipe wall. Depending on the rate of change of temperature, errors in the order of μm could result. It can be concluded that very advanced temperature compensation methods would be required to reliably track thickness changes below the μm mark using ultrasonic techniques.

4 Effects of Corrosion-Induced Surface Morphology Changes

In this section, we are interested in the effect that uneven surface morphology changes have on ultrasonically monitored thicknesses. Corrosion is a complicated phenomenon and is a large field of study. It is a degradation mechanism that can result in thickness loss of many forms. The loss can be spatially uniform as in etching or spatially nonuniform as in pitting corrosion and any combination of the two depending on the material and environmental conditions. It is therefore important to note that every component will have spatial variation of its thickness due to the nature of the corrosion process. Furthermore, it is very simplistic to condense the resulting surface condition into a single thickness value that results from an ultrasonic thickness gauging measurement. It is the end user of the information who needs to decide what metric is important to him, does he/she require the mean, max, or minimum thickness?

In corrosion engineering, it is common to use the weight loss of coupons or electrochemical potential or current measurements to infer a corrosion rate. When doing these measurements, one inherently averages over the area of the component, and this effectively results in a mean wall thickness loss evaluation. In the field, it is therefore accepted that some form of mean or average is reported when quoting a corrosion rate. For the purpose of this work, it was therefore decided to compare ultrasonic wall loss trend measurements on rough surfaces to the mean wall thickness change.

Gajdacsi et al. [5, 7] investigated the effect of surfaces morphology changes on the error in ultrasonically evaluated corrosion rate compared to the actual underlying corrosion rate by means of simulation. They simulated the evolution of many hundreds of random rough surfaces with the same statistical properties, simulated ultrasonic signals reflected from these surfaces, and used different signal processing algorithms to estimate the wall loss trend/corrosion rate from the ultrasonic data. They concluded that standard signal processing algorithms resulted in rather large corrosion rate error estimates and developed a new signal processing algorithm that they termed adaptive cross-correlation (AXC). This new algorithm performed much better and was able to monitor corrosion rates (90% of simulated population) to within −10 to +25% of the actual trend of the mean wall loss for spatially random and evolving thickness loss (Gaussian rough surfaces of RMS 100–300 μm and correlation length 1 mm). This means that if the actual underlying mean wall loss of the rough surface was 1 mm/year, in 90% of cases, the algorithm would report a rate in between 0.75 and 1.1 mm/year. These figures were quoted for the worst-case surface conditions and the algorithm did perform better on flat surface morphologies and evolutions. Figure 6 summarizes the trend error results for the different algorithms (first arrival [FA], peak-to-peak [P2P], cross-correlation [XC] and

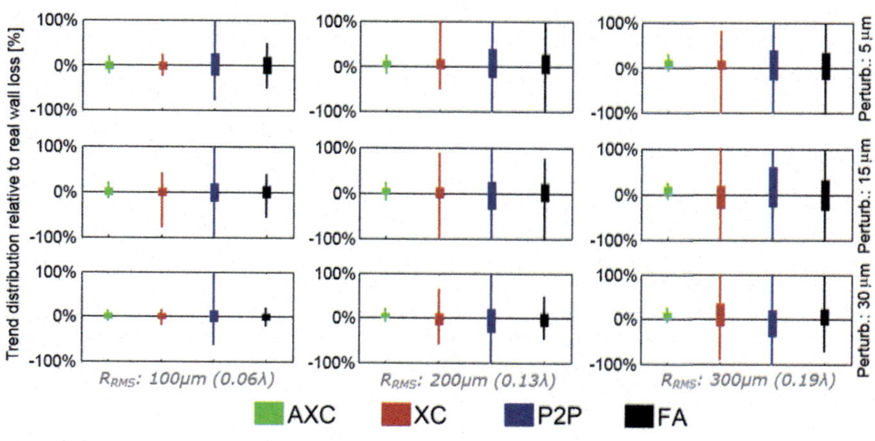

Fig. 6 Trend error distributions as computed by Gajdacsi [5, 7] on simulated ultrasonic monitoring data from rough evolving surfaces. The results show the performance of different signal processing algorithms

adaptive cross-correlation [AXC]) and surface morphology evolutions. The interested reader is referred to [7] for more detail.

5 Conclusions

This paper summarized current knowledge of the performance of a state-of-the-art wireless ultrasonic corrosion monitoring system. These new systems make it possible to collect ultrasonic wall thickness data with unprecedented frequency and precision. It was discussed that temperature compensation and accurate material data is required to make very precise wall thickness measurements. However, it was also discussed that temperature compensation with independent thermocouple measurement data to sub-μm precision will be very difficult in the field because of the existence of temperature gradients within pipe walls during heating and cooling processes. The topic of corrosion-induced surface morphology changes on ultrasonically monitored data was also introduced. The work by Gajdacsi et al. on advanced signal processing algorithms (AXC) to mitigate the uncertainties that are introduced by the changes in surface morphology was recalled. Gajdacsi et al. concluded that corrosion rates on spatially uniform corroding surfaces could be track to within −10 to +25%.

References

1. ASTM Standard E797/E797M-10 (2010) Standard practice for measuring thickness by manual ultrasonic pulse-echo contact method. ASTM International, West Conshohocken, PA. DOI:10.1520/E0797_E0797M-10
2. EN14127:2011 (2011) Non-destructive testing. Ultrasonic thickness measurement. European Standard, CEN, Brussels
3. www.permasense.com. Last accessed 14 August 2015
4. Yi W-G, Lee M-R, Lee J-H, Lee S-H (2006) A study on the ultrasonic thickness measurement of wall thinned pipe in nuclear power plants. In: 12th Asia-Pacific Conference on NDT, Auckland, New Zealand
5. Gajdacsi A (2015) High accuracy ultrasonic degradation monitoring. PhD thesis, Mechanical Engineering Department, Imperial College London
6. Gajdacsi A, Cegla F (2013) High accuracy wall thickness loss monitoring. In: Review of Progress in Quantitative Nondestructive Evaluation, AIP Conference Proceedings 1581, American Institute of Physics, Melville, NY, pp. 1687–1694
7. Gajdacsi A, Cegla F (2015) Ultrasonic wall loss monitoring of rough surfaces. In: Review of Progress in Quantitative Nondestructive Evaluation, AIP Conference Proceedings 1650, American Institute of Physics, Melville, NY, pp. 856–862

Passive SHM System for Corrosion Detection by Guided Wave Tomography

Tom Druet, Bastien Chapuis, Manfred Jules, Guillaume Laffont and Emmanuel Moulin

Abstract Guided elastic waves emitted by a sensor and propagating to another one are often used as the physical way of detecting the defect in Structural Health Monitoring (SHM) systems of plate-like structures. However, the implementation of SHM systems is restricted in many situations by the necessity to store or to harvest the electric energy necessary to emit the waves. A promising way to overcome this constraint is to use techniques based on the cross-correlations of the ambient noise in place in the structure. The idea is to take advantage of the elastic noise naturally present in the structure (due to engine vibrations or aero-acoustic turbulences on the fuselage of an aircraft for example) in order to avoid the emission of the elastic waves by the SHM system. The complexity of the embedded SHM system is therefore reduced. We present here studies of noise cross-correlation techniques that have been conducted with the aim of doing passive guided wave tomography of extended defects (such as corrosion) using an array of piezoelectric (PZT) transducers. Noise is generated by spraying compressed air on the surface of a thin aluminum plate. Passive measurements are compared to active signals to demonstrate the effectiveness of the cross-correlation technique. Experimental results which come from tomographic time-of-flight imaging algorithms will also be described. Finally, an extension of this technique using Fiber Bragg Gratings (FBG) optic sensors will be presented.

Keywords NDT · Nondestructive testing · Structural health monitoring
SHM · Tomography

V-1670-15

T. Druet · M. Jules · G. Laffont
CEA LIST, Gif-sur-Yvette, France

B. Chapuis (✉)
NDE Department, CEA LIST, Gif-sur-Yvette, France
e-mail: bastien.chapuis@cea.fr

E. Moulin
Université de Valenciennes et du Hainaut-Cambrésis, Valenciennes, France

B. Chapuis and E. Sjerve (eds.), *Sensors, Algorithms and Applications for Structural Health Monitoring*, IIW Collection, https://doi.org/10.1007/978-3-319-69233-3_3

1 Introduction

The Structural Health Monitoring (SHM) consists of embedding sensors in a structure such as an aircraft or a naval ship in order to detect defects (for example, cracks or corrosion in metallic materials or delamination in composite materials) before a serious fault occurs in the structure. Guided elastic waves emitted by a sensor and propagating to another one are often used as the physical way to detect defects. In aeronautics, the classical approach generally aims at minimizing the number of sensors to limit the embedded mass as well as the sensors intrusiveness within the structure. Comparisons between current signals and baseline signals are often performed in order to reveal the presence of defects [1]. However, this method may not be robust under certain conditions such as changes in temperature, stress, sensors aging, etc.

A possible strategy to avoid the use of baseline signals can consist in increasing the number of sensors and performing guided wave tomography. Indeed, more relevant physical information is obtained from the structure, making the diagnosis more robust. Moreover, tomography algorithms produce images that are much easier to interpret than temporal signals. Implementation of SHM systems with a large number of sensors could be intrusive with the use of piezoelectric transducers. Optical fiber sensors using Fiber Bragg gratings (FBGs) for dynamic strain measurements should allow multiplexing capabilities with low intrusiveness in the structure. However, FBGs are generally used not only as sensors but also as a source of elastic waves. A promising way to tackle these constraints is to use techniques based on cross-correlation of the ambient acoustic noise present in the structure. It has been shown that, under certain conditions, transient responses between two sensors can be passively estimated from the cross-correlation of ambient noise [2–4]. The idea is to take advantage of the elastic noise naturally present in the structure (due to engine vibrations or aero-acoustic turbulence on the fuselage of an aircraft for example) in order to avoid the need for emission of the elastic waves by the SHM system.

This paper shows an active tomography image which has been performed with a time-of-flight tomography algorithm using experimental data produced by piezoelectric transducers. Furthermore, a comparison between active signals and passive signals which comes from cross-correlation of ambient noise produced by spraying compressed air on an aluminum plate shows that it is possible to detect time-of-flight passively. In Sect. 2, the tomography algorithm used for this work is presented. Then, in Sect. 3, the experimental setup used to acquire the data necessary to obtain the tomography images is described. Finally, in Sects. 4 and 5, the experiments are presented showing that it is possible to use cross-correlation of ambient noise present in the plate-like structure to get the data required for the tomography algorithm. Several configurations of sensors were studied using piezoelectric transducers, Fiber Bragg Gratings, and a combination thereof.

2 Time-of-Flight Tomography Algorithm

Images shown in this paper are obtained with a time-of-flight tomography algorithm which uses the Simultaneous Iterative Reconstructive Technique (SIRT) [5]. Straight ray assumption is taken within this framework. Straight ray tomography does not take into account refraction and diffraction. By ignoring diffraction, only defects bigger than the first Fresnel zone [6] and varying slowly are correctly reconstructed. By ignoring refraction, the algorithm is limited to low-contrast flaws. Better algorithms that take into account refraction and diffraction exist [7] and will be studied in future work.

A simulation respecting these assumptions (i.e., no refraction and diffraction) was conducted (see Fig. 1). The configuration in Fig. 1a shows 30 sensors (which are depicted by yellow points) and three different defects. The image values matrix has a size of 23 × 23 pixels. Input data is a set of time-of-flight values deduced from dispersion curves of the A_0 mode at 30 kHz propagating in a 2-mm-thick plate. The absolute group velocity is 1447 m/s when waves propagate in the healthy part of the plate and 1300, 1200, and 1056 m/s when they propagate in defect zones (i.e., zones of reduced thickness which can illustrate corrosion phenomenon). In practice, image smoothing is interesting since corrosion and pixels will unlikely be superimposed on each other. That is why Fig. 1 and the experimental image in Fig. 3b are smoothed by performing interpolation.

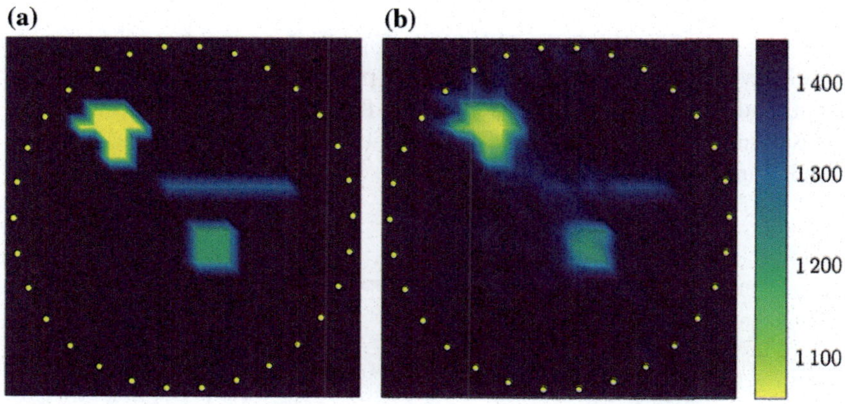

Fig. 1 **a** Configuration; **b** Result after 1 iteration

3 Experimental Setup

The time-of-flight algorithm described in the previous section helped, among others things, to design an experimental setup (see Fig. 2). The setup is composed of a computer running LabView which controls an oscilloscope and a multiplexer, a generator which is synchronized with the oscilloscope by a trigger, an amplifier, a filter, and finally the piezoelectric transducers attached to the 2–mm-thick aluminum plate. Phenyl salicylate (SALOL) is used to bond localized thin calibrated aluminum layers on the plate. For experimental purposes, these layers will serve as an easily removable defect with a somewhat similar effect than corrosion on wave velocity due to thickness change.

4 Active Experimental Image

Figure 3 shows an experimental active absolute time-of-flight tomography image (Fig. 3b) which results from the algorithm described in Sect. 2. In practice, reversible flaws (Fig. 3a) were used by adding a thin aluminum plate glued (with SALOL) on the plate to be inspected. The term "Absolute tomography" means that data only comes from current signals. This way, the method should be quite robust as baseline signals are never used. Input data was obtained by emitting a 1.5 cycle tone-burst at 177.5 kHz to the piezoelectric transducers. Time-of-flights from first S_0 wave packet were identified by the algorithm for each pair of sensors. The two flaws of Fig. 3a are easily distinguishable in Fig. 3b. Experimentally, there were local adhesive disbonds between aluminum plates, zones with poor SALOL adhesion, and fluctuations in the thickness of the SALOL. All of those reasons partly explain why the reconstructed defects do not fit exactly the real lines of rectangular and circular aluminum plates.

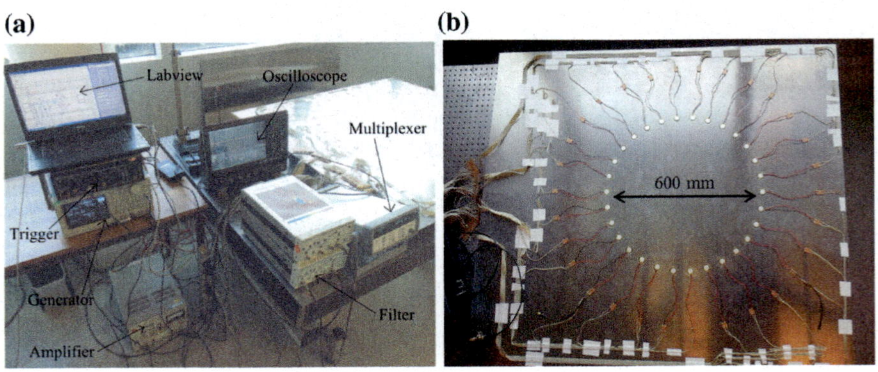

Fig. 2 **a** Experimental setup; **b** 30 piezoelectric transducers (Ø 18 mm) stuck on the 2-mm-thick aluminum plate

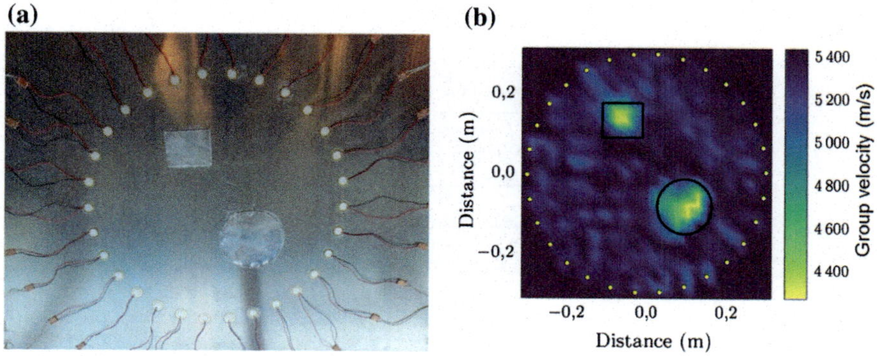

Fig. 3 **a** Experimental configuration; **b** Absolute time-of-flight tomography (smoothed velocity map)

5 Passive Experimental Results

The input data necessary to obtain the tomography images is a set of time-of-flight values, as presented in previous sections. This section aims at assessing the retrieval of this kind of data from passive signals. Equation 1 shows the cross-correlation formulation with $\vec{u}$ the displacement field at the point $\vec{x}$ and $\vec{v}$ the displacement field at the point $\overrightarrow{x'}$. Every passive signal obtained in this section results from cross-correlation of ambient noise recorded during the amount of time necessary (10 s in our experiments) for it to converge toward Green's functions:

$$\underline{\underline{C}}\vec{u},\vec{v}\left(t,\vec{x},\overrightarrow{x'}\right)=\lim_{T\to\infty}\frac{1}{T}\int_0^T\vec{u}(\tau,\vec{x})\vec{v}\left(t+\tau,\overrightarrow{x'}\right)^T d\tau \tag{1}$$

5.1 *Piezoelectric Sensors (PZT)*

A comparison between active and passive signals obtained, thanks to cross-correlation of ambient elastic noise, is presented in Fig. 4.

Eight cycles tone-burst at 20 kHz were generated in order to obtain the active signals shown in Fig. 4. The elastic noise used for the passive signals was produced by spraying compressed air on the aluminum plate surface. The absolute values of those signals are plotted in Fig. 4 for several distances between sensors. This way, the propagation of the first wave packet is shown for each active signal. Passive signals show a satisfactory reconstruction of Green's functions at least for the first wave packet. Indeed, the result is symmetrical (because of acoustic reciprocity between the two observation points) and first wave packets are adequately superimposed on passive and active signals. It is therefore possible to get time-of-flight from passive signals. This is a reason to think that passive tomography is

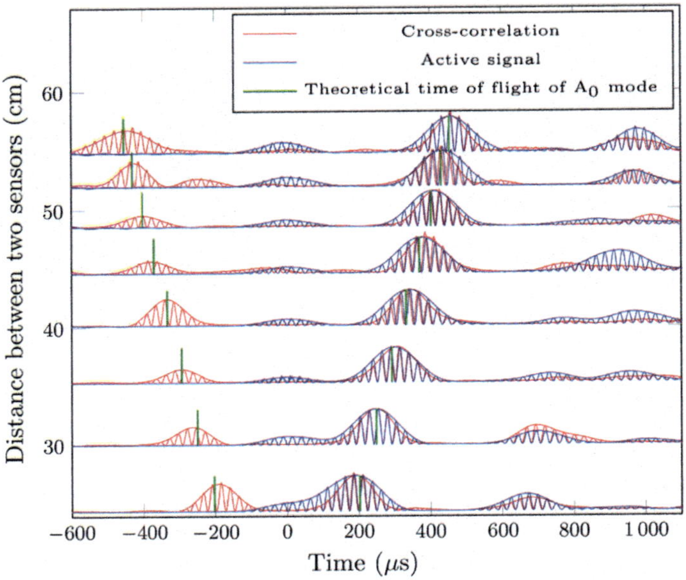

Fig. 4 Comparison between active and cross-correlation signals

Fig. 5 Reconstruction of passive signals for high frequencies—theoretical time-of-flights for S_0 and A_0 at 250 kHz

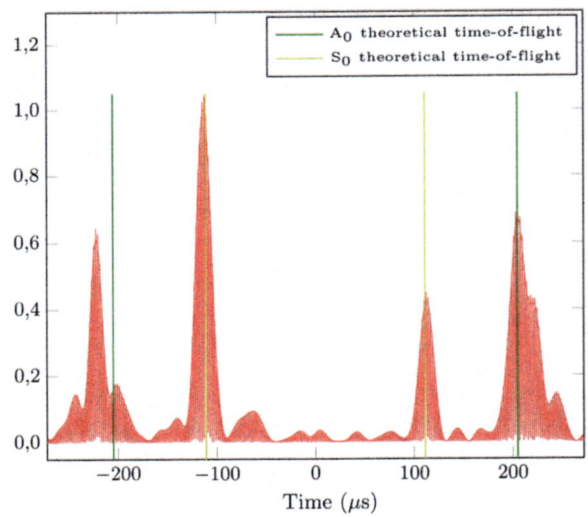

promising. We can notice in the passive signals some parasite wavelets caused by the spatial distribution of the ambient noise [8, 9]. Indeed, the more the ambient structural noise is spatially uniformly distributed, the better the convergence toward Green's functions will be.

Figure 5 shows that it is possible to obtain time-of-flight on passive signals for higher frequencies, of at least 250 kHz (for the ambient noise created by spraying the air on the aluminum plate). Here, the S_0 as well as A_0 modes can be detected in Fig. 5 but it is not possible to correctly identify the time-of-flight for A_0. Indeed, S_0 reflections arrive at the same time as the first A_0 wave packet.

5.2 *Fiber Bragg Gratings (FBG)*

It has already been demonstrated that FBGs are able to detect guided waves emitted by piezoelectric transducers [10, 11]. Here, we perform passive experimental measurements using only fiber Bragg gratings sensors. With such passive measurements, PZT transducers for elastic wave emission would no longer be necessary, reducing the intrusiveness of the sensors in the structure.

Elastic noise is generated in the structure using the same setup as in previous sections: by spraying compressed air on the aluminum plate surface. Figure 6 shows the passive reconstruction of the A_0 mode at 40 kHz between FBGs bonded on the surface of a 2-mm-thick aluminum plate and spaced 400 mm apart from each other.

This demonstrates the feasibility of passive reconstruction using FBGs only. Indeed, the theoretical time-of-flight values perfectly match the maximum of the first wave packet (A_0 mode). Figure 7 shows the passive reconstruction (in red) between an FBG and a PZT for several distances between both sensors. The PZT is used here only as sensor of the elastic noise, not to emit the guided waves. It has been chosen for convenience in this experiment because it is easier to move than the FBG. It confirms that the first wave packet is well reconstructed and that it is possible to correctly identify its time-of-flight with passive measurements using

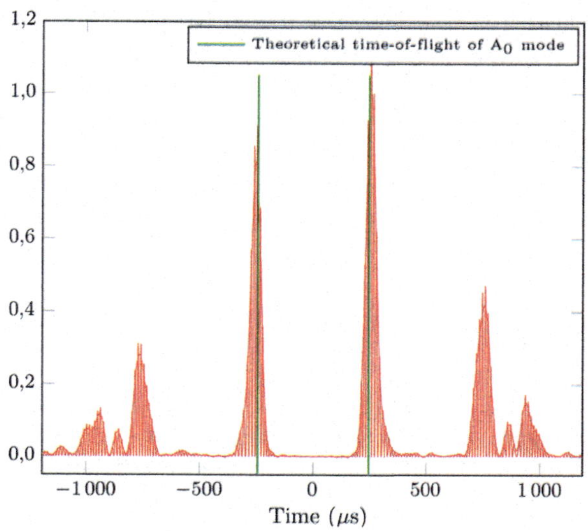

Fig. 6 Reconstruction of passive signals with two fiber Bragg gratings bonded on the surface of a 2-mm-thick aluminum plate

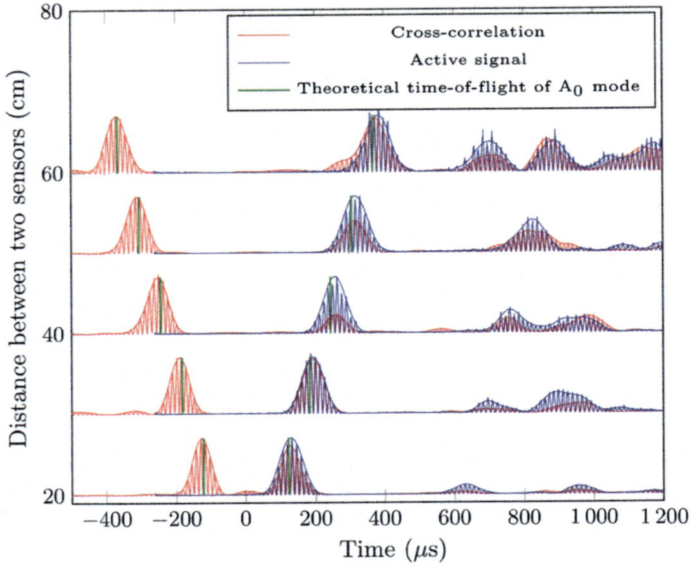

Fig. 7 Reconstruction of passive signals between a piezoelectric transducer and a fiber Bragg grating. Comparison between active and cross-correlation signals

only FBG. In Fig. 7, in the same way as in Fig. 4, we can see that the first wave packets are adequately superimposed on passive and active signals.

6 Conclusion

This paper has shown that it is possible to image defects experimentally by using piezoelectric transducers and a time-of-flight tomography algorithm without the need of baseline signals. Active experimental tomography using guided wave data was performed in this paper. It was also shown that it is possible to detect time-of-flight by using cross-correlation of ambient noise with piezoelectric transducers as well as with fiber Bragg gratings. Passive experimental tomography is the next step of our study by merging the results obtained in these two domains.

References

1. Croxford A, Wilcox P, Drinkwater B, Konstantinidis G (2007) Strategies for guided-wave structural health monitoring. Proc R Soc A 463:2961–2981
2. Weaver RL, Lobkis OI (2001) Ultrasonics without a source: thermal fluctuation correlations at MHz frequencies. Phys Rev Lett 87(13):134301

3. Sabra KG, Winkel ES, Bourgoyne DA, Elbing BR, Ceccio SL, Perlin M, Dowling DR (2007) Using cross correlations of turbulent flow-induced ambient vibrations to estimate the structural impulse response. Application to structural health monitoring. J Acoust Soc Am 121(4):1987–1995
4. Moulin E, Abou Leyla N, Assaad J, Grondel S (2009) Applicability of acoustic noise correlation to structural health monitoring in non-diffuse field conditions. Appl Phys Lett 95:094104
5. Kak A, Slaney M (1988) Principles of computerized tomographic imaging. IEEE Press, New York, p 275
6. Williamson PR (1991) A guide to the limits of resolution imposed by scattering in ray tomography. Geophysics 56(2):202–207
7. Huthwaite P, Simonetti F (2013) High-resolution guided wave tomography. Wave Motion 50:979–993
8. Chehami L, Moulin E, De Rosny J, Prada C, Assaad J, Benmeddour F (2015) Experimental study of passive defect detection and localization in thin plates from noise correlation. In: Proceedings of ICU 2015. Metz, France
9. Colombi A, Boschi L, Roux P, Campillo M (2014) Green's function retrieval through cross-correlations in a two-dimensional complex reverberating medium. J Acoust Soc Am 135:1034–1043
10. Betz D, Thursby G, Culshaw B, Staszewski W (2003) Acousto-ultrasonic sensing using fiber Bragg gratings. Smart Mater Struct 12:122–128
11. Botsev Y, Arad E, Tur M, Kressel I, Ben-Simon U, Gail S, Osmont D (2008) Structural health monitoring using an embedded PZT-FBG ultrasonic sensor array. In: Proceedings of the fourth European workshop on structural health monitoring. Cracow, Poland

On the Way to Use Welds for Monitoring of Structures Integrity

Slah Yaacoubi, Daniel Chauveau, Fethi Dahmene and Mahjoub El Mountassir

Abstract This chapter deals with the development of a novel magnetostrictive sensor/actuator that is used to generate ultrasonic guided waves. Instead of attaching a sensor to the structure to generate guided waves, this describes using part of the structure to be monitored as the sensor in the ideal case, or adding a joint-weld or deposit in the less than ideal case. Results obtained via this new development are presented, and compared with findings obtained through use of a commercial magnetostrictive sensor.

Keywords Welds · Guide waves · Magnetostrictive sensor

1 Introduction

Screening piping structures via ultrasonic guided waves (UGW) nondestructive testing and monitoring (NDT&M) techniques are maturing. In the current structural health monitoring context of piping structures, the sensor/actuator should be attached permanently to the structure that is being monitored. This makes the monitoring expensive since the cost (including qualification exercises) of the sensor/actuator is expensive.

The weld itself, or a part of the structure, might be used as the passive part of an alternative sensor. A new sensor/actuator prototype was developed at the Institut de Soudure. This sensor exploits the magnetostrictive effect of the weld and/or the structure, provided that special processes are used during the manufacturing stage to enhance their magnetostrictive properties. Trials using this prototype have been

---

V-1671-15

S. Yaacoubi (✉) · F. Dahmene · M. El Mountassir
Institut de Soudure, Plateforme RDI CND, Yutz, France
e-mail: s.yaacoubi@isgroupe.com

D. Chauveau
Innovation Department, Institut de Soudure Industrie, Villepinte, France

B. Chapuis and E. Sjerve (eds.), *Sensors, Algorithms and Applications for Structural Health Monitoring*, IIW Collection, https://doi.org/10.1007/978-3-319-69233-3_4

carried out, and generation and detection of waves have been performed successfully. Experimental results using this prototype sensor will be presented and compared with results acquired using a commercial actuator/sensor [1]. The result of detection of a relatively small defect using this prototype will also be shown.

2 Magnetostriction, Magnetostrictive Actuator, and Sensor

Magnetostriction is a phenomenon which exists mainly in ferromagnetic materials such as iron, nickel, cobalt, and their alloys (other materials exist). The principle of magnetostriction is based on the magneto-mechanical properties of these materials. When this kind of material is placed in a magnetic field, it creates an elastic deformation. This physical phenomenon is due to the existence of many small elementary magnets which constitute the ferromagnetic material or equivalent.

The magnetostrictive sensing and actuating technology generates and detects elastic waves electromagnetically. These elastic waves are created through a pulsed alternating current flowing in a coil, which generates a magnetic field. This magnetic field is combined with the permanent magnetic field in the material. This phenomenon is called the direct effect or the Joule effect. After their propagation, these elastic waves will be reflected by reflectors (defects, welds, ends, etc.) and propagate back until they reach the receiver coil. Reciprocally, this mechanical deformation (i.e., these elastic waves) leads a change in magnetic properties of the environment. This is the inverse magnetostrictive effect (or Villari effect).

A conventional magnetostrictive sensor used to perform guided wave inspection consists of two parts: a ferromagnetic strip adhesively bonded to the structure and a coil. This has the significant advantage of being very compact. Due to its small thickness, the sensor is straightforwardly usable in areas that have poor accessibility [2]. Nevertheless, for structural health monitoring purposes, the bonding of this strip needs to be qualified with the same lifetime of the structure. The aim of this work is to replace the function of this strip by a weld or a deposit. Some examples are shown in the next section.

3 Weld or Deposit for Generating and Receiving Waves

3.1 Propagation in Healthy Tube

As explained previously, welds or deposits can generate and receive guided waves via the magnetostrictive effect, provided that they are manufactured with a material which is able to be deformable. This material should not be expensive, and it should also not affect the weldability or durability of the weld. In addition, a specific process is needed to improve the magnetostrictive properties of the selected material.

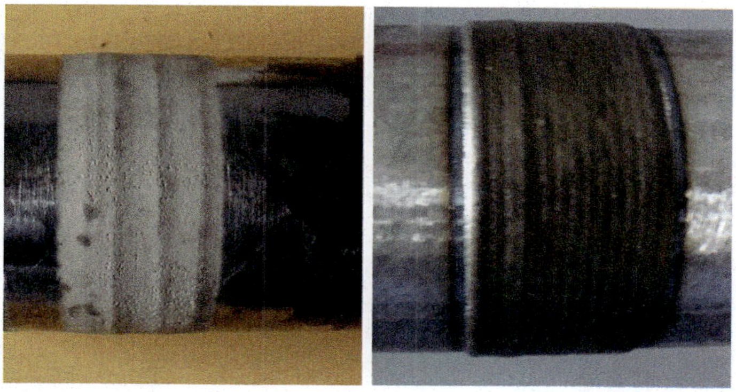

Fig. 1 Examples of prototypes

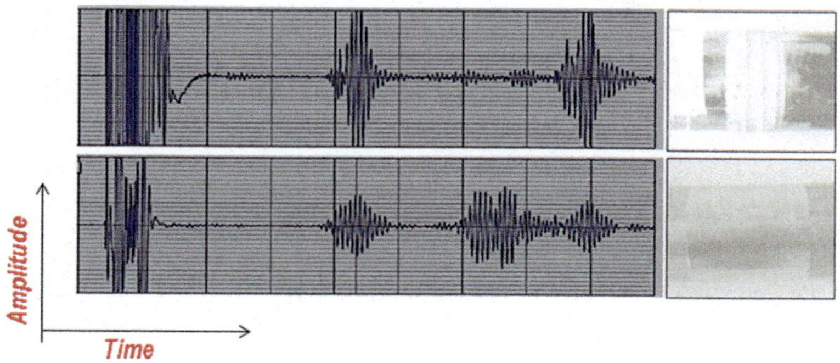

Fig. 2 Example of results obtained by the developed prototype (top) and a commercial transducer (bottom)

Different types of materials and processes and/or deposits have been tested (one patent is granted and two are pending). Figure 1 shows two examples of realization obtained with two different methods.

Figure 2 shows waveforms obtained using the pulse-echo measurement configuration. This figure shows a comparison between results obtained through the developed prototype in the current study (top) and a commercial strip sensor (bottom), with different gains. The main echoes that can be seen in this figure correspond to the end of pipe, which was used for these experiments. As it can be seen, the signal-to-noise ratio is comparable. This experiment, which has a promising result, now has to be performed in more attenuating pipes.

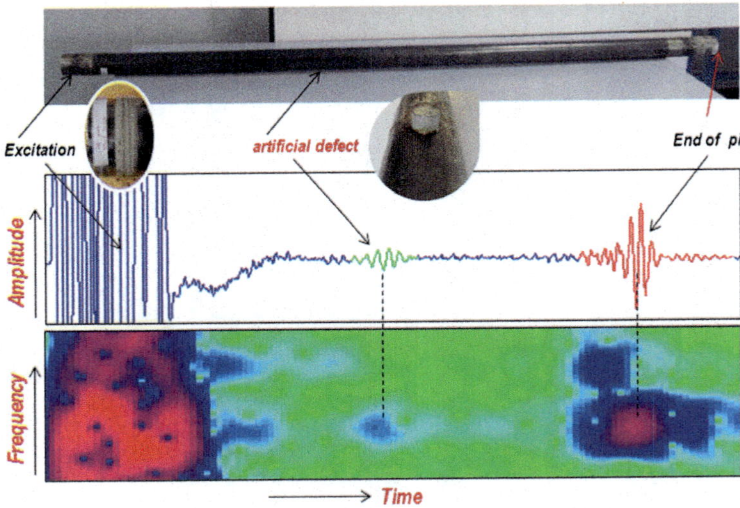

Fig. 3 Pipe used for experiment, artificial defect, and result obtained shows that detection was performed successfully

3.2 Defect Detection

To test the detection sensitivity of the current prototype for relatively small defects, an artificial defect was created on the external surface of the pipe by attaching a magnet to it. Figure 3 shows the pipe used in this experiment and the result that was obtained. The green and red echoes correspond, respectively, to the defect and the end of the pipe. As it can be seen in the figure, defect detection was successful and the signal-to-noise ratio is high.

4 Conclusion

A novel magnetostrictive sensor/actuator prototype is presented. Basing on this prototype, guided wave experimental trials were performed in a pipe with no defects and the collected result was compared to a result obtained using a commercial strip sensor. The prototype was then used to detect an artificial defect having a small cross-sectional area. The results of this work are promising. Future experiments will be performed in pipes with various attenuation coefficients [3] and different defects with various characteristics to further investigate this new development in guided wave generation and reception.

References

1. http://www.gwanalysis.com/mss.html
2. Yaacoubi S, et al. (2012) Contrôle des structures longues via les ondes guidées: application aux pipelines et tuyauteries. Downloadable from http://www.isgroupe.com/fr/metiers/centre-technique-rd/documentation/stc/TechnicalArchives/Contr%C3%B4le%20des%20structures%20longues%20via%20les%20ondes%20guid%C3%A9es.pdf
3. Leinov E, Lowe MJS, Cawley P (2015) Investigation of guided wave propagation and attenuation in pipe buried in sand. J Sound Vibr 347:96–114

Use of Guided Wave Inspections to Monitor the Integrity of Nuclear Power Station Boilers

Peter Mudge

Abstract Owing to potential concerns about creep occurring in the mid-part of the central tubular spines of nuclear plant boilers, as a precautionary measure TWI and its subsidiary, Plant Integrity Ltd, were contracted in 2002/2003 to develop a test procedure to check the integrity of the spines using their Teletest ultrasonic guided wave testing (GWT) system. At that time, both the thickness and complexity of the component were beyond the normal operating window of the Teletest equipment, which was designed for testing pipelines. To validate the operation and capabilities of the method, a full-scale mock-up component was produced. Field testing was commenced in 2003. The eight boilers around a reactor were tested when it was shut down for a statutory outage, which occurs on a 3-year cycle for each reactor.

Periodic tests have been carried out on all 32 spines at statutory outages since 2003. Up to 2013, all spines produced consistent responses. However, in September 2013, one spine at Heysham 1 Power Station, Reactor 1 produced a significantly different response. The application of guided wave testing enabled a significant failure in a nuclear power plant to be avoided.

- Investigation of the factors affecting the responses from flaws in the affected region enabled a capability to be established and this contributed to the production of a successful safety case for return to service of the plants affected.
- Special tooling and data gathering equipment have enabled continued monitoring.
- GWT is now an integral part of the ongoing assessment of integrity of the spines.

Keywords Guided waves · Nuclear power plant · Boiler · Welds

V-1672-15

P. Mudge (✉)
TWI Ltd., Cambridge, UK
e-mail: peter.mudge@twi.co.uk

B. Chapuis and E. Sjerve (eds.), *Sensors, Algorithms and Applications for Structural Health Monitoring*, IIW Collection, https://doi.org/10.1007/978-3-319-69233-3_5

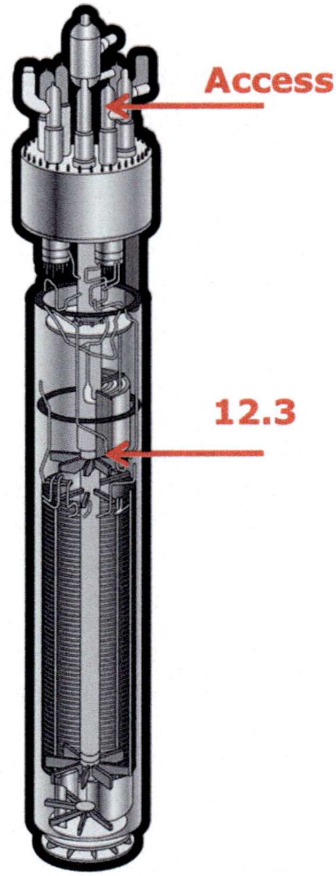

Fig. 1 Schematic of a 'pod' boiler, showing the central spine and the weld 12.3, of special interest

1 Background

Hartlepool and Heysham 1 power stations in the UK are nuclear plants of the advanced gas-cooled reactor (AGR) design, currently operated by EDF Energy. The coolant, which passes directly from the reactor core through the boilers, is CO2. Both plants began supplying the national grid in 1983 and are of 1210 and 1150 MW capacity, respectively. Between them, they supply around 7% of the current UK electricity demand. Each station consists of two reactors, each of which is surrounded by eight boilers of a unique 'pod' design. Other AGR stations in the UK have boilers of a different design.

A pod boiler is shown schematically in Fig. 1. The 'pods' were designed to facilitate replacement of the boilers but this has never been attempted. A major design feature is that the boiler internals are suspended from a central tubular spine, 520 mm in diameter, up to 60 mm thick and around 21 m long, which is supported at the top of the boiler. This is not pressurised but the boiler feedwater, which enters at the top of the spine via a nozzle, is carried down to the bottom of the column in an array of tubes inside it. The flow of feedwater keeps the top of the spine at around 160 °C in operation, whereas the central part may reach over 500 °C. The spines are made from a combination of creep-resisting alloys in the hottest section, stainless steel, and carbon steel. There are also a number of attachments at the mid- and end points from which the boiler tubes are suspended. Only the top 2 m of the spine is accessible above the boiler closure, as may be seen from Fig. 1; Most of the length of the spine is completely inaccessible for inspection.

2 Inspection Issues

Owing to potential concerns about creep occurring in the mid-part of the spines, as a precautionary measure TWI and its subsidiary, Plant Integrity Ltd were contracted in 2002/2003 to develop a test procedure to check the integrity of the spines using their Teletest ultrasonic guided wave testing (GWT) system [1], there being no other viable nondestructive testing method which could be applied to the spines while they were in an operational condition. Of main concern was the condition of the so-called '12.3' weld located at the top of the forging at the mid-point where radial arms were attached to support the boiler tubes. At that time, both the thickness and complexity of the component were beyond the normal operating window of the Teletest equipment, which was designed for testing pipelines. To validate the operation and capabilities of the method, a full-scale mock-up component was produced. The main criterion for the inspections was to determine that the signal received from the far end of the spine could be observed, and this would therefore infer that the spine was intact. Field testing was commenced in 2003. The eight boilers around a reactor were tested when it was shutdown for a statutory outage, which occurs on a 3-year cycle for each reactor. Access for inspection was not only restricted to a short length of the spine column accessible above the top of the boiler closure, but the spine is also surrounded by other superheater headers and structural restraints which greatly restrict access. Figure 2 shows an original Teletest guided wave tool mounted on the upper spine in the location indicated.

Periodic tests have been carried out on all 32 spines at statutory outages since 2003. Up to 2013, all produced consistent responses with the spine end signal clearly visible. However, in September 2013, one spine at Heysham 1 Power Station, Reactor 1 produced a significantly different response from the 12.3 weld in one of the hottest parts of the boiler, with the spine end signal not visible. As stated, at the time the plant was on a statutory outage. Even during such an outage, only very limited confirmatory NDT could be carried out via remote visual inspection.

3 Inspection Findings

Figure 3 shows the inspection data from 2013 compared with the previous inspection, carried out in 2010. The so-called 12.3 weld is at 10.4 m from the transducer tool. Initial follow-up investigations centred on efforts to establish if this change was due to factors other than from a flaw in the spine, either arising from the

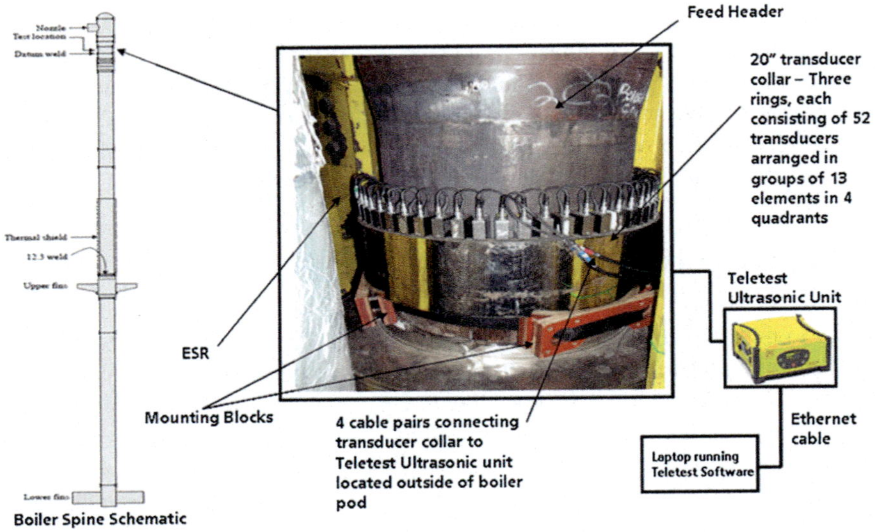

Fig. 2 Teletest guided wave tool mounted on a boiler spine. The main area of potential concern was the '12.3' weld, around halfway down the column, adjacent to the upper radial arm attachments

test equipment or from operational conditions in the boiler. No evidence was found for this, so efforts turned to establishing the relationship between Teletest responses and potential flaw size. The full-scale mock-up, produced for the 2002 developments, was used for experimental studies. In addition, the numerical modelling group at TWI was engaged to prepare FEA (finite element analysis) models of both the mock-up (which was manufactured from carbon steel only) and the real component and to run simulations of the ultrasonic inspections, again to establish the likely size of flaw that could have produced the observed response.

4 Investigation of Inspection Performance

As there was no realistic alternative to Teletest for the examination of the boiler spines, extensive further studies were undertaken to determine:

- Whether there was any evidence of flaws in the 12.3 weld in any of the remaining 31 boiler spines from previous data;
- The detection capability of Teletest for flaws at the 12.3 weld, and at other locations along the spines;
- Factors that affect the flaw detection performance;
- Analysis procedures for Teletest data which are more rigorous than a simple observation of a response from the far end of the spine;
- Whether the performance was sufficient to support a safety case for return to service of the plants, in conjunction with other measures being undertaken by

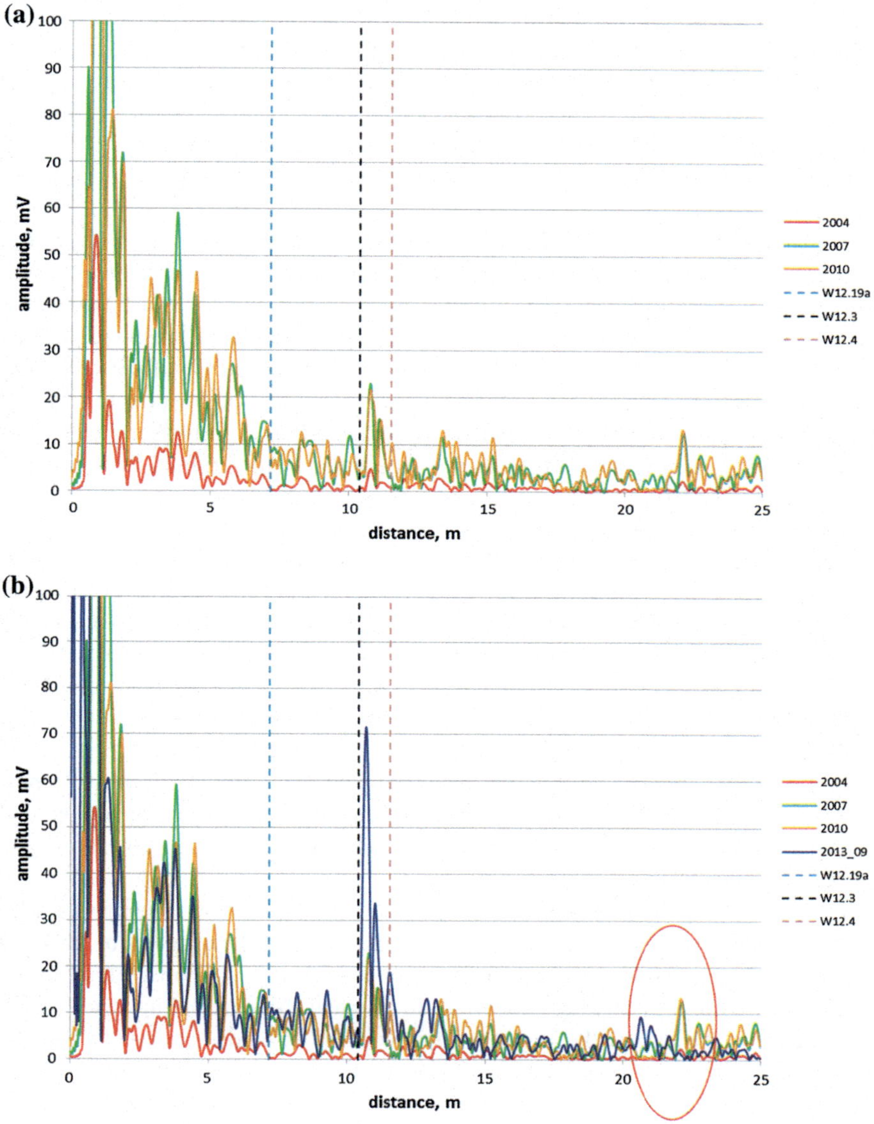

Fig. 3 Responses from the affected boiler spine at 10.4 m range from **a** 2010 and **b** 2013. There is a ~10 dB increase in the signal at 10.4 m range and the end of spine signal at ~22 m has disappeared

EDF energy. (NB the boilers for each reactor are operated in pairs (quadrants) so that the quadrant affected by the flaw was isolated and has not operated since September 2013).

In parallel with this, there was a requirement for enhanced Teletest equipment to facilitate more frequent inspections in-plant to check the condition of the spines. Access to the spine to perform the tests, as shown in Fig. 2, is only possible when a considerable amount of equipment and insulation has been removed from the area around the top of the boiler. While this is done at major statutory outages, this is not feasible when the plant is in operation or at short outages, such as for refuelling. A requirement to obtain Teletest data frequently led to the development of tooling that could remain in-situ while the plant was operating and linked to an external connector so that data could be collected rapidly at refuelling outages and even during operation.

A major survey was undertaken of all the Teletest data collected on the spines since commencement of testing in 2003. During the procedure development, it was determined that the torsional T(0, 1) wave mode should be used and that the best test frequency was 36 kHz, based on the need to propagate the ultrasound through the 60-mm-thick sections where the radial arms are attached. While 36 kHz was used for routine reporting, a range of test frequencies was used for field testing and this additional information in the archive was examined in case further evidence of cracking was present in those other results. Additionally, the original reporting was only based on the T(0, 1) response from the spine end. In the review, the flexural F (1, 2) responses were also taken into consideration.

The outcome of this study was a more extensive set of criteria for evaluation of the Teletest data from the spines. It was determined that there was no evidence of the presence of any flaws in any other spines comparable to that detected in the Heysham spine. However, at this stage, there was no quantitative information about the limits of capability of the technique for the detection of planar defects at the 12.3 location. Therefore, a significant programme was set up using both FEA modelling and practical testing to establish the performance limits.

Full 3-D FEA models of both the real component and the mock-up were produced which allowed a parametric study of the guided wave generation and propagation along the spine and the interaction with flaws. Figure 4 shows the model of the actual spine.

Of greatest significance was the geometry immediately surrounding the 12.3 weld. This is a circumferential weld which connects the 35-mm-thick tubular to the top end of the central forging which carries the radial arm supports. This forging flares out to 60 mm thick where the arms are attached. An external upstand 'nib' is machined into the end of the forging, onto which is welded a concentric thermal shield, which may be seen in Fig. 4. The geometry is shown schematically in Fig. 5. Of particular relevance is the spacing between the 'nib' feature and the weld. The distance between the top of the nib and the weld centreline is 63 mm, while the wavelength of the T(0, 1) mode at 36 kHz is 90.5 mm. Not only does this mean that there is a response from this region, even if no flaw is present, but also any flaws at the weld cannot be resolved in time from the characteristic signal from the geometry.

In parallel with the modelling studies, an experimental programme was commenced on the full-scale mock-up to demonstrate the performance of the Teletest system and tooling and to validate the models. Aspects of this are shown in Fig. 6.

Both the models and the experimental tests produced an unexpected result. The presence of the 'nib' upstand in close proximity to the planar flaw resulted in an

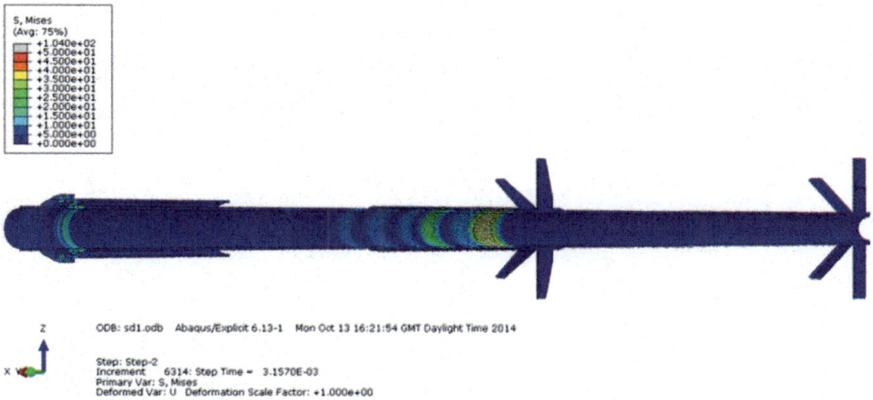

Fig. 4 FEA model of the boiler spine showing the main geometric features: the radial support arms, the thermal shrouds, the support 'boss' at the top of the spine and the simulation of the Teletest tool

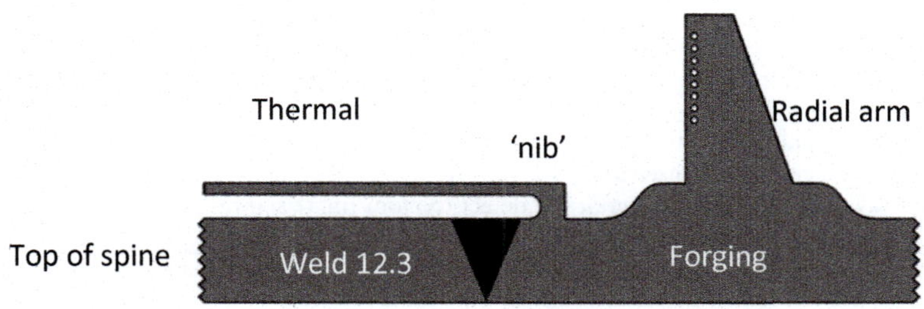

Fig. 5 Geometry surrounding the 12.3 weld. The wall of the tubular is 35 mm, while the forging increases to 60 mm at its thickest point

interference effect, such that there was no observable response from the through-thickness defect until it was around 20% of the annular cross section of the wall of the tubular spine. Figure 7 shows a comparison of the model and experimental results. This differs from the accepted linear relationship between the area of a through-thickness notch as a proportion of the annulus and reflectivity [2]. As stated, there is a reflection from the local geometry, even in the absence of any flaw, so that the data do not pass through the origin in Fig. 7. The response reaches a minimum at 15% cross-sectional area (CSA). Further, the amplitude does not increase above that initial value until the defect occupies approximately 30% of the circumference. For a 6 dB increase above the defect-free response, Fig. 7 suggests that a defect would need to be in excess of 50% CSA to be reliably detected. This would be an unacceptably low level of sensitivity to enable any ability of the inspection to provide any forewarning of failure.

The drop in response as a flaw is introduced results from an interference effect between the geometric 'nib' and the plane of the flaw. The resultant signal is highly sensitive to the separation between the two. A study of the effect of lateral position

Fig. 6 Experimental trials on the full-scale mock-up at TWI. The Teletest tool at the far end and the through-thickness notch inserted at the 12.3 weld centreline are shown. The thermal shroud was cut away to allow the insertion of the defect (bottom right) and re-welded to the 'nib', also visible, at each stage

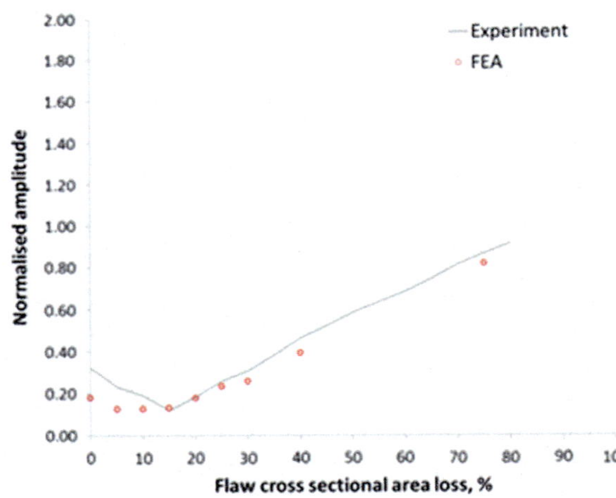

Fig. 7 Responses from a through-thickness, part-circumferential notch at the 12.3 weld from 36 kHz guided wave tests

of the flaw relative to the weld centreline showed that there was a sinusoidal relationship between position and response amplitude. Figure 8 shows this effect for two cases: a fully circumferential notch of 41.7% CSA and a through-thickness notch of 26.1% CSA. The responses both follow a sinusoidal variation with position but are offset from each other.

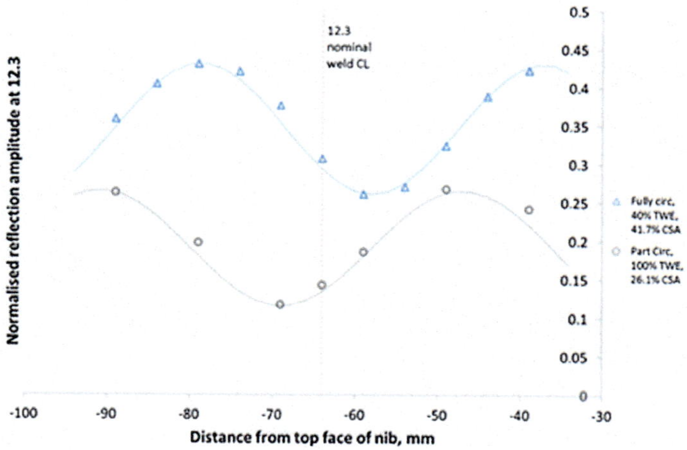

Fig. 8 Predicted FEA responses from a fully circumferential, part-wall notch and a part-circumferential, through-wall notch as a function of axial position

The weld centreline position is near the minimum for both types of defect, confirming the interference behaviour shown in Fig. 7. Local examination of the crack itself showed that it lay in the lower HAZ of the weld, 48 mm above the ‘nib’. It may be seen that this is at a maximum of the response curve for a fully penetrating defect. Further, the cracking mechanism is much more likely to occur in the HAZ than in the weld metal, so that it was possible to establish a reliable limit of detection for such defects of 23% CSA in the upper and lower HAZ of the 12.3 weld. This information was used as part of the safety case for return to service of all four reactors in late 2014.

5 Tooling and Monitoring

A requirement of the safety case for return to service was to perform more detailed Teletest inspections more often. To allow transducer tools to remain on the spines while in service to facilitate testing, even at brief outages, a high temperature permanently installed (HTP) tool was developed, Fig. 9. HTP tools were installed on all operational spines and have enabled the inspection programme to be implemented during continued plant operation.

A further mitigating measure was to introduce insulating material between the shroud and the spine to reduce surface temperature. The HTP tools permitted guided wave data to be gathered while the plants are in operation to check the efficacy of this and also to offer the possibility of online condition monitoring. Automated Teletest data logging devices were developed (Fig. 9b) to allow this to be done round the clock.

(a) (b)

Fig. 9 **a** HTP Teletest tool at installation, without protective cover. **b** Automated Teletest data logger device for capture of data while the plant is operating

6 Concluding Remarks

The application of guided wave testing enabled a significant failure in a nuclear power plant to be avoided.

- Investigation of the factors affecting the responses from flaws in the affected region enabled a capability to be established and this contributed to the production of a successful safety case for return to service of the plants affected.
- Special tooling and data gathering equipment have enabled continued monitoring.
- GWT is now an integral part of the on-going assessment of integrity of the spines.

Acknowledgements The authors would like to thank other members of the project team: Nick Farrant, Jane Allwright, Adam Wilkinson, Sina Fateri, Barry Elborn, Harvey Whitmore, and Jonathan Hunt for their contributions. Thanks are also due to EDF Energy Nuclear Generation for their permission to publish this work.

References

1. Mudge PJ (2004) Practical enhancements achievable in long range ultrasonic testing by exploiting the properties of guided waves. In: Proceedings of 16th WCNDT, Montreal, Canada, 30 Aug–3 Sep 2004
2. Demma A, Cawley P, Lowe M, Roosenbrand AG (2003) The reflection of the fundamental torsional mode from cracks and notches in pipes. J Acoustical Soc Am 114(2)

A Study on Prediction of Fatigue Crack Propagation Life Using MFC Sensors

D. J. Oh, J. H. Lee, J. M. Lee and M. H. Kim

Abstract One of the most important tasks in offshore and ship structures is to avoid structural failure due to defects induced by cyclic loading and structural deterioration. While the detection technology for such flaws is essential in order to ensure the integrity of structures, it is difficult to identify and measure such defects in real time with sufficient reliability. In this regard, various structural health monitoring (SHM) techniques have been developed for detecting defects and damage from many research communities. One of the key parameters in structural integrity assessment (SIA) is a precise estimation of the stress intensity factor (SIF). The SIF is used for estimating the fatigue crack propagation life based on linear elastic fracture mechanics (LEFM). In this study, the SIF of a simple test specimen is measured in real time using a macro-fiber composite (MFC) sensor. MFC sensor is a high-performance smart material used both for actuator and sensor offering good flexibility and reliability. This study examines the feasibility of MFC sensors for an SHM application with particular attention to fatigue crack monitoring. Finally, the fatigue crack propagation life calculated by an MFC sensor is compared with experimental results.

Keywords Structural health monitoring · Structural integrity assessment
MFC sensor · Stress intensity factor · Fatigue crack growth life

V-1673-15 (XIII-2585-15).

D. J. Oh · J. H. Lee · J. M. Lee · M. H. Kim (✉)
Department of Naval Architecture and Ocean Engineering,
Pusan National University, Busan, South Korea
e-mail: kimm@pusan.ac.kr

B. Chapuis and E. Sjerve (eds.), *Sensors, Algorithms and Applications for Structural Health Monitoring*, IIW Collection, https://doi.org/10.1007/978-3-319-69233-3_6

1 Introduction

When the existing defects or flaws are exposed to cyclic loading and deterioration, they grow at various rates and cause structural damage. While short-term or small flaws do not cause a structural failure, long-term or large flaws may significantly deteriorate the structural performance. This status is referred to as structural failure. In particular, damage can accumulate over a long period of time associated with corrosion and fatigue loading. Structural damage can also result from ship collisions, aircraft landings, and earthquakes. In this respect, structural health monitoring (SHM) technology that can evaluate damage qualitatively and quantitatively is required.

The main functions of SHM are to evaluate damages and the remaining service life during operation period of structures. During the total service life, SHM provides a diagnosis for the state of the structure. In particular, these technologies can be quantitatively used to evaluate damage and to identify the failure location. As shown in Fig. 1, an SHM system typically consists of three categories: detecting damage or flaws, recording the cumulative damage, and providing a diagnosis of the structure. For damage detection, the expected damage mechanism should first be determined and then matched to the available sensors. In particular, sensors for monitoring damage in real time should have sufficient durability and accuracy under various environmental conditions. Then, the appropriate data, such as the accumulation of damage and load history, is collected using a data acquisition system. The signal data obtained by the integrity monitoring system, in parallel with the previously registered data, is then used for a diagnosis of the structure. The information from the integrity monitoring system with the knowledge based on fracture mechanics makes it possible to determine the prognosis and the health management of the structure.

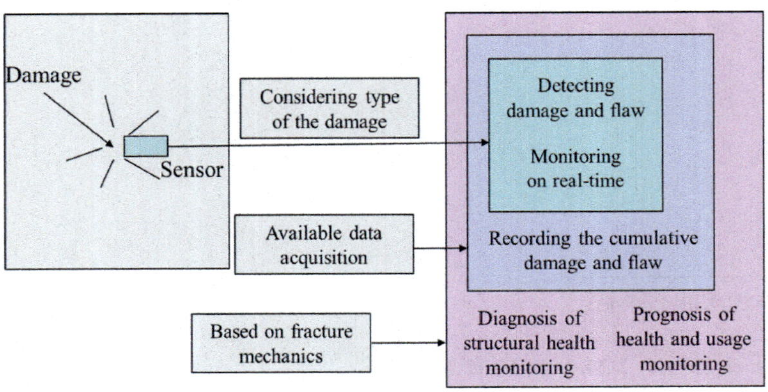

Fig. 1 Structure of the typical SHM system

Various SHM systems have been developed in order to evaluate the effect of defects and structural damage [1]. In general, traditional nondestructive evaluation (NDE) techniques, such as ultrasonic technology, acoustic emission, magnetic field analysis, and X-ray analysis, are typically used to examine for flaws in the structure. However, these techniques are readily available for real-time detection of flaws. In this regard, online detecting systems mounted in situ on structures are being developed using smart materials. Sun et al. [2] and Chaudhry et al. [3] evaluated structural integrity based on an impedance-based health monitoring system using piezoelectric materials. They demonstrated that piezoelectric patches are successful in crack and damage detection.

Fujimoto et al. [4] successfully measured the SIF of two-dimensional cracks in a structure using piezoelectric patches. Piezoelectric materials possess a property that generates electric charge on its surface when stress is applied. Based on the piezoelectric constitutive law and fracture mechanics, the SIFs of Mode I as well as of Mode II are calculated. In addition, Butrym et al. [5] proposed a combined method to simultaneously detect fatigue crack initiation and fatigue crack growth using the impedance method and MFC sensors for SIF measurement. They demonstrated that the measured SIF obtained by MFC sensors can predict the remaining service life of an aluminum beam.

Sumant and Maiti [6] proposed a method to detect the size and location of an edge crack in a beam using PZT patches attached to the top and bottom edges of the beam. They performed numerical simulations for evaluating the influence of two crack parameters, size and location. Based on the numerical results, they measured the fatigue crack length according to the change of the output voltage in the PZT patches. Ryles et al. [7] described fatigue crack detection in metallic structures. They used nonlinear acoustics and Lamb wave-based inspection for fatigue crack detection and monitoring. Staszewski et al. [8] also demonstrated that Lamb waves and a 3D laser vibrometer are able to detect fatigue cracks. Zagrai and Giurgiutiu [9] described the application of the EM impedance method for damage detection in thin circular plates. Zhang et al. [10] presented an approach to detect fatigue crack length and location by measuring a set of the electric potential difference, or resistance values, at several points on the coating sensor. They used nickel–phosphorus coating covered on the structural surface as sensors.

In this study, an approach based on macro-fiber composite (MFC) sensors is proposed that can observe crack initiation and predict the crack propagation life in real time. The stress intensity factor (SIF) is essential for calculating the fatigue crack propagation life based on linear elastic fracture mechanics (LEFM). In this study, MFC sensors are employed for direct measurement of SIF in the presence of fatigue cracks. A pair of MFC sensors is attached in the vicinity of a crack tip for real-time measurement of the SIF. The experimental SIF values are compared with the analytical SIF values. Finally, the prediction of fatigue crack propagation life, which is obtained using experimental SIF values, is compared with actual propagation life.

2 Prediction of Fatigue Crack Growth Life Using SIF

Fracture of the structures may occur at stress levels below the yield strength when cracks are present in the structure. For certain locations where cracks are difficult to avoid, a special parameter should be used to evaluate a possible failure of the structure. In the context of LEFM, the SIF is useful for evaluating the failure of structure once a fatigue crack occurs. The SIF is mainly used to accurately predict the stress state near the crack tip. In particular, the magnitude of the SIF depends on various parameters, such as crack geometry, size, and location of the crack and applied load. When the SIF exceeds the critical stress intensity factor (K_C) associated with the fatigue crack growth, the structure under consideration reaches a final failure.

The fatigue life is commonly divided into two stages: crack initiation life and crack propagation life. The crack initiation life is the number of cycles required to induce a fatigue crack in region I. The fatigue crack in region I is defined as an initiation of microcracks at the material level. The crack propagation life is the number of cycles required to induce a final fracture. The fatigue crack propagation life can be predicted based on an empirical equation established by a relationship between the SIF range and the fatigue crack growth rate (d*a*/d*N*). In the region II, a simple equation called the Paris law is commonly used to describe the fatigue crack propagation life [11]:

$$\frac{da}{dN} = C(\Delta K)^m \tag{1}$$

where ΔK is the stress intensity factor range, and da/dN is the fatigue crack growth rate. Here, C is the intercept of the vertical axis, and m is the slope of the da/dN–ΔK curve that can be identified by interpolating the fatigue crack growth data in region II. The SIF equation for single-edge-notched tension (SENT) and center crack tension (CCT) specimens are given by

$$\Delta K = F \cdot \Delta\sigma\sqrt{\pi a} \tag{2}$$

$$F_{\mathrm{SENT}} = \frac{1.1\left(1 - 0.2\left(\frac{a}{W}\right)\right)}{1 - \left(\frac{a}{W}\right)^{3/2}} \tag{3}$$

$$F_{\mathrm{CCT}} = \sqrt{\sec\frac{\pi a}{w}}\left[1 - 0.025\cdot\left(\frac{2a}{w}\right)^2 + 0.06\cdot\left(\frac{2a}{w}\right)^4\right] \tag{4}$$

where $\Delta\sigma$ is the stress range remote from the crack and a is the crack size. The geometry parameters F_{SENT} and F_{CCT} are presented according to type of the specimen by Ashby and Tada [12, 13]. In Eqs. (3) and (4), w is the width of the test specimen.

Then, the fatigue crack propagation life can be obtained by integrating Eq. (1).

$$N_f = \int_0^{N_f} dN = \int_{a_i}^{a_f} \frac{da}{C(\Delta\sigma)^m (\pi a)^{m/2} F^m} = \frac{1}{C(\Delta\sigma)^m (\pi)^{m/2} F^m} \int_{a_i}^{a_f} \frac{da}{a^{m/2}} \quad (5)$$

In Eq. (5) a_i is the initial crack length, and a_f is the final crack length.

3 Estimation of Stress Intensity Factor Using the MFC Sensors

An MFC sensor consists of rectangular piezoceramic rods sandwiched between layers of adhesive, electrodes, and polyimide film as shown in Fig. 2. The electrodes are attached to the film in an interdigitated pattern which transfers the applied voltage directly to and from the ribbon-shaped rods. As a thin, surface conformable sheet it can be applied (normally bonded) to various types of structures or embedded in a composite structure. The MFC sensor is a low-profile actuator and sensor offering high performance, durability, and flexibility in comparison with other piezoelectric materials [14].

Piezoelectric material induces an electric charge on its surface by the applied stress. Fujimoto demonstrated that it is possible to measure SIFs directly using piezoelectric material [4]. According to the method described by Fujimoto, SIF is directly measured in real time based on the output voltage of the MFC sensors. The

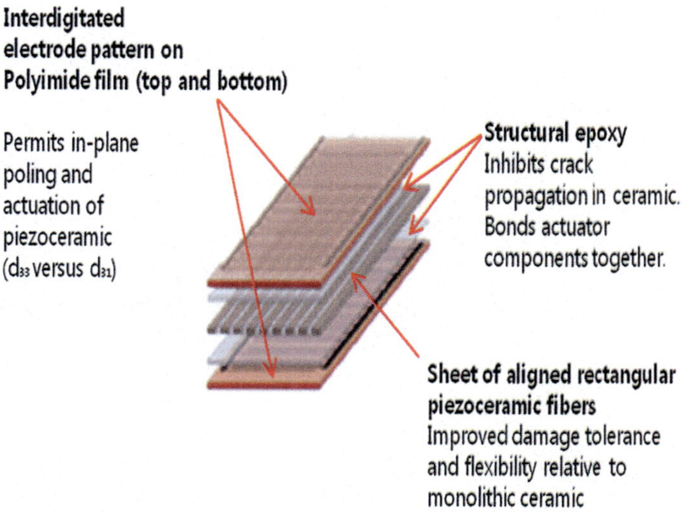

Fig. 2 Structure of an MFC sensor

output voltage V is recorded by an oscilloscope from MFC sensors attached on the surface of the specimen. In addition, the peak-to-peak voltage of each MFC sensor is saved along with the fatigue crack growth. The relationship between output voltage V and stress is given by

$$V = C \cdot K \int (\sigma_x + \sigma_y) \mathrm{d}S \tag{6}$$

where C (V/N) is a coefficient determined by the properties of the MFC sensor and the adhesive condition, and K is a coefficient related to the stress condition. Poisson's ratio is $v = 0.3$, and K is 1 for the plane stress condition and 0.74 for the plane strain condition, respectively.

Based on the LEFM, the stress sum ($\sigma x + \sigma y$) of the two-dimensional through-thickness cracks at the crack tip is given as follows [8]:

$$\sigma_x + \sigma_y = \frac{2K_{\mathrm{I}}}{\sqrt{2\pi r}} \cos\frac{\theta}{2} - \frac{2K_{\mathrm{II}}}{\sqrt{2\pi r}} \sin\frac{\theta}{2} \tag{7}$$

Substituting Eq. (7) into Eq. (6), the output voltages can be expressed as

$$V_A = C \cdot K \int_{H_0}^{H+H_0} \int_{-L_1}^{L_2} \frac{2K_{\mathrm{I}}}{\sqrt{2\pi r}} \cos\frac{\theta}{2} - \frac{2K_{\mathrm{II}}}{\sqrt{2\pi r}} \sin\frac{\theta}{2} \mathrm{d}x\mathrm{d}y \tag{8}$$

$$V_B = C \cdot K \int_{-H-H_0}^{H_0} \int_{-L_1}^{L_2} \frac{2K_{\mathrm{I}}}{\sqrt{2\pi r}} \cos\frac{\theta}{2} - \frac{2K_{\mathrm{II}}}{\sqrt{2\pi r}} \sin\frac{\theta}{2} \mathrm{d}x\mathrm{d}y \tag{9}$$

where r is the crack length and θ is the crack angle. After the output voltages V_A and V_B are measured, the SIFs can be calculated based on the following equations:

$$K_{\mathrm{I}} = \frac{V_A + V_B}{2CKG_1} \tag{10}$$

$$K_{\mathrm{II}} = \frac{V_B - V_A}{2CKG_2} \tag{11}$$

In Eqs. (10) and (11), parameters G_1 and G_2 are represented as follows:

$$G_1 = \int_{H_0}^{H+H_0} \int_{0}^{L_2} \sqrt{\frac{\sqrt{x^2 + y^2} + x}{\pi(x^2 + y^2)}} \mathrm{d}x\mathrm{d}y \tag{12}$$

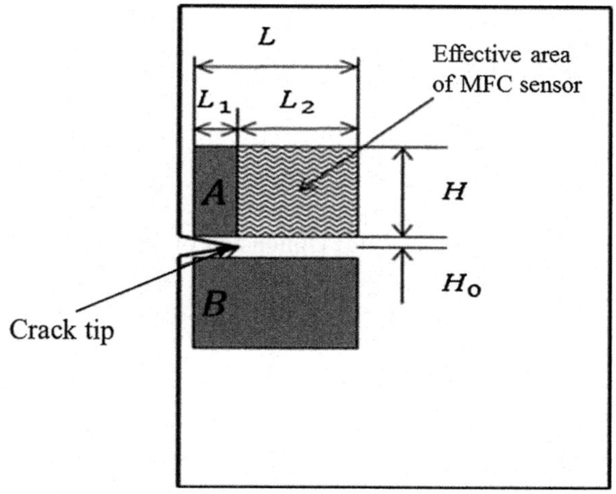

Fig. 3 Effective area of the MFC sensor with increasing fatigue crack

$$G_2 = \int_{H_0}^{H+H_0} \int_{0}^{L_2} \sqrt{\frac{\sqrt{x^2+y^2}-x}{\pi(x^2+y^2)}}\,\mathrm{d}x\mathrm{d}y \tag{13}$$

In general, the effective area of the MFC changes with the increase of the fatigue crack length [5]. Considering the origin located at the crack tip, Eqs. (12) and (13) are integrated from *0* to L_2 in the direction of the fatigue crack growth. As the fatigue crack grows, the integration range changes with the decrease of the effective area of the MFC. Detailed integration range of the effective area is illustrated in Fig. 3.

4 Experimental Procedure

In this study, test specimens made of mild steel are prepared for the fatigue crack growth tests. The yield and tensile strengths of the test material are 220 and 350 MPa, respectively. In order to consider the crack location, two types of specimens are considered: single-edge-notched tension (SENT) and center crack tension (CCT) specimens. Each specimen has a machined notch at the midspan of the specimen. In the SENT specimen, a notch of length 3 mm is inserted at an edge of the specimen. In the CCT specimen, a machined center notch of length $2a$ = 40 mm is inserted at the center of the specimen. Detailed dimension and geometry of the test specimens are illustrated in Fig. 4.

As shown in Fig. 5, the MFC sensor employed in this study had an active area of L = 28 mm, H = 14 mm, and the thickness 0.3 mm. The detailed dimensions and properties of the MFC sensors are summarized in Table 1.

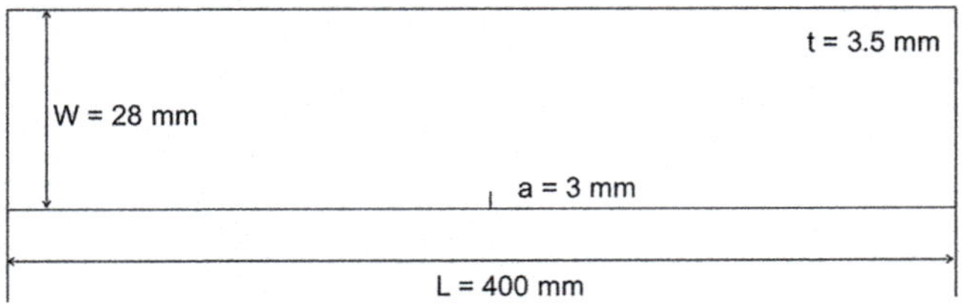

(a) Dimension of the SENT specimen

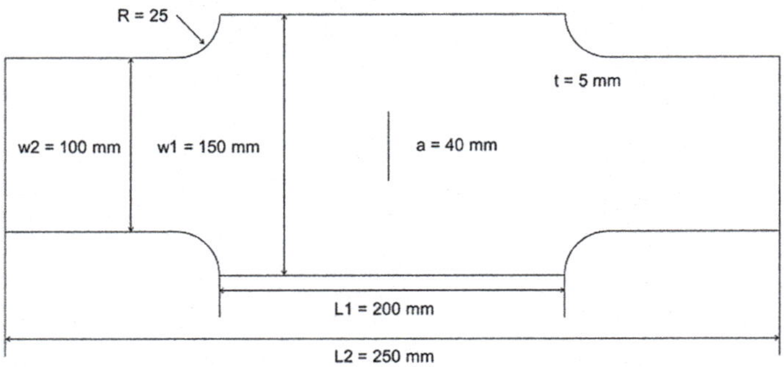

(b) Dimension of the CCT specimen

Fig. 4 Detail dimension of the test specimens

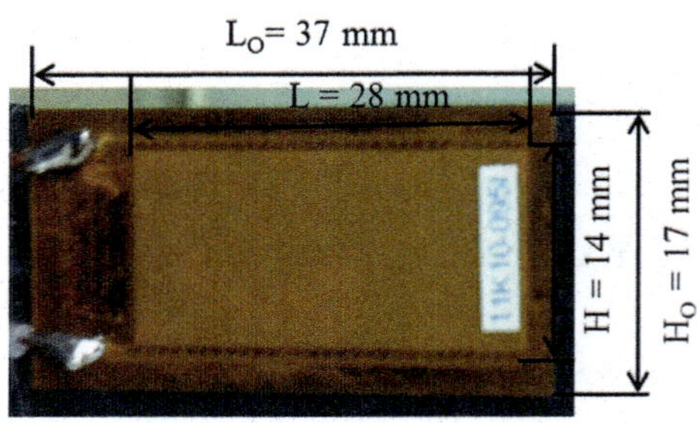

Fig. 5 Active area of the MFC sensor

Table 1 Dimension and properties of the MFC sensor

Overall dimension (L_o, H_o)	37 mm × 17 mm
Active area (L, H)	28 mm × 14 mm
IDE spacing	0.5 mm
Capacitance	1.2 nF
PZT type	Navy type II
Max voltage	− 500 to + 1500 V
Max tensile strain	4500 ppm
Thickness	0.3 mm

Fig. 6 Experimental setup

A series of fatigue crack growth tests were conducted using a servohydraulic test machine with a maximum load capacity of ± 20 ton (SAGINOMIYA). The stress range was 117.5 MPa for the SENT specimen and the CCT specimen using a sinusoidal loading. Stress ratio and test frequency were 0.1 and 3–5 Hz, respectively.

The MFC patches were attached on both sides of the crack tip using epoxy glue. The distance between each MFC sensor and the crack tip was 3 mm. The change in the effective area of the MFC patch along the direction of fatigue crack growth is considered to be the integration area. An experimental setup with the attached MFC sensors is shown in Fig. 6. The peak-to-peak voltage outputs are measured using a four-channel digital oscilloscope equipped with USB output to a laptop computer. The output voltage of the MFC sensors obtained in real time is later used for calculating the SIFs of the test specimens.

The calibration constant C varies with the type of MFC sensor and the adhesion condition in the test specimen. Therefore, the MFC sensors should be calibrated to obtain a calibration constant C before each fatigue test. The constant C is calculated based on the measurement of the peak-to-peak voltage at the applied stress range. The equation of the calculating calibration constant C is given by

$$C = \frac{V}{\Delta\sigma \times S} \tag{14}$$

where V is the peak-to-peak voltage output, $\Delta\sigma$ is the applied stress range, and S is the active area of the MFC sensor. The constant C from this experiment is obtained as 0.005 for the SENT specimen and 0.004 for the CCT specimen, respectively.

5 Test Results

During the fatigue test, the changes of the peak-to-peak voltage and the fatigue crack length were measured periodically. A microscope was employed for the measurement of the fatigue crack length. The fatigue test was terminated when the fatigue crack reached a length of the MFC sensor.

For the SENT specimen, the fatigue crack length corresponding to the number of cycles is shown in Fig. 7a. In addition, the plots of the output voltage (V_A, V_B) versus the fatigue crack length and the number of cycles are presented in Fig. 7b, c, respectively. It is clearly seen that the output voltage decreases as the fatigue crack grows. In particular, we observed that the SENT specimen loses its capacity when the fatigue crack length reaches 4–6 mm in 280,000 cycles. And then, the output voltage quickly decreases when the fatigue crack grows beyond the crack length of 3–4 mm. The test results of the CCT specimen are shown in Fig. 8. Similar to the results of the SENT specimen, the output voltage gradually decreases with the increase of the

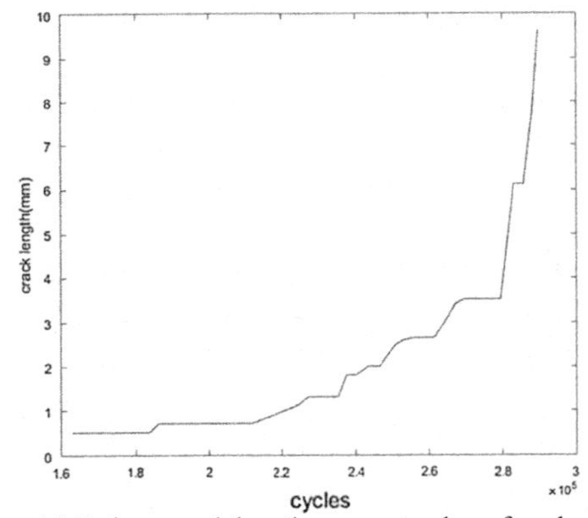

(a) Fatigue crack length versus number of cycles

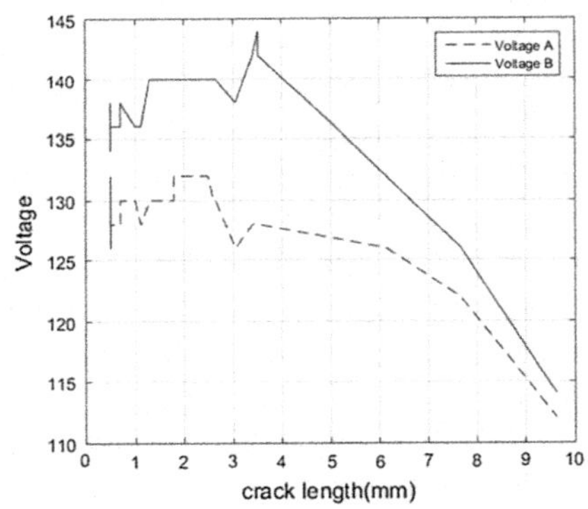

(b) Output voltage versus fatigue crack length

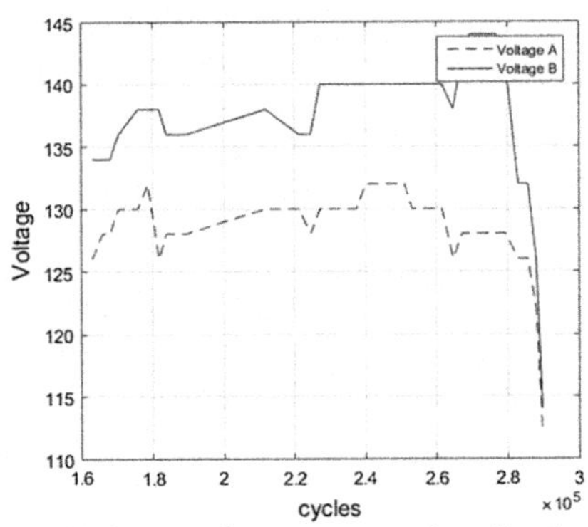

(c) Output voltage versus number of cycles

Fig. 7 Results of the SENT specimen

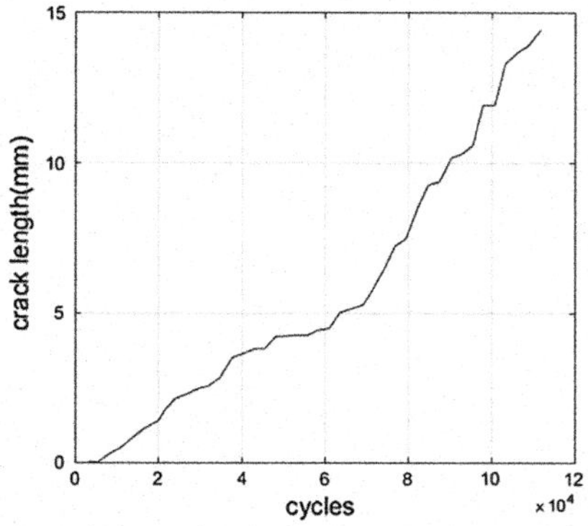

(a) Fatigue crack length versus number of cycles

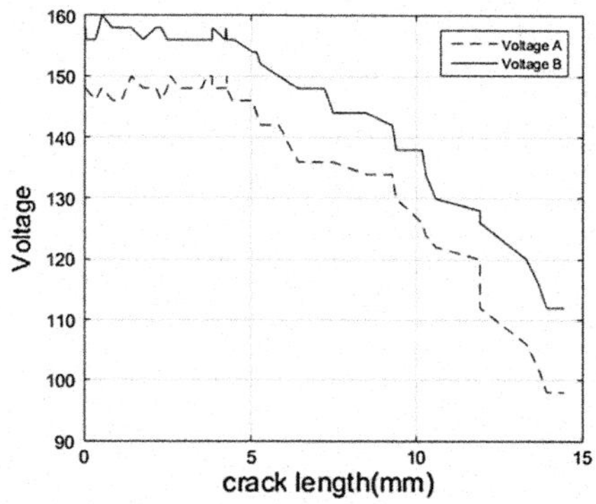

(b) Output voltage versus fatigue crack length

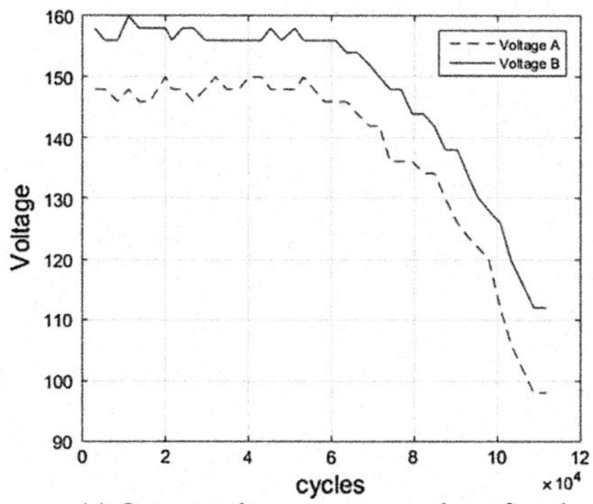

(c) Output voltage versus number of cycles

Fig. 8 Results of the CCT specimen

fatigue crack length. In particular, the slope of a–N curve increases drastically when the fatigue crack length reaches 5–6 mm. Based on these observations, the output voltage measurement can be used as an indication of the fatigue crack growth.

As shown in Fig. 9, the SIFs obtained by the analytical method based on LEFM are compared with the measured SIFs from experiments. In the case of the SENT specimen, in particular, a large error of the maximum 40% between experimental SIF and analytical SIF are observed when the fatigue crack length increases in a rapid manner. For the CCT specimen, in contrast, the difference between the experimental SIF and analytical SIF is small when the fatigue crack grows. It should be noted that the difference is significant at the early stage of fatigue crack

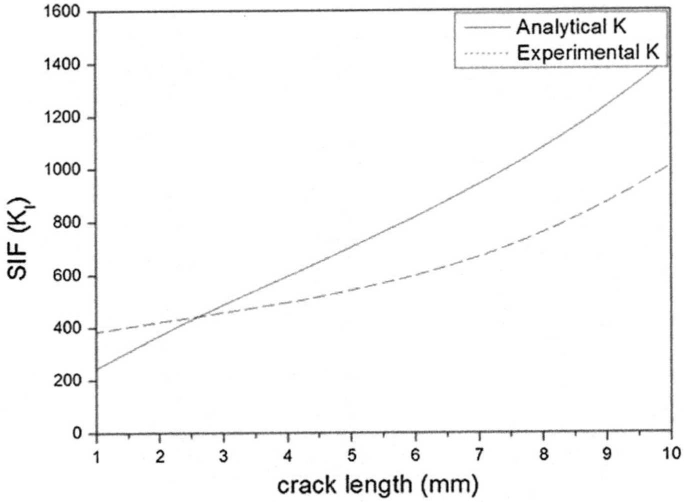

(a) SIF versus fatigue crack length of the SENT specimen

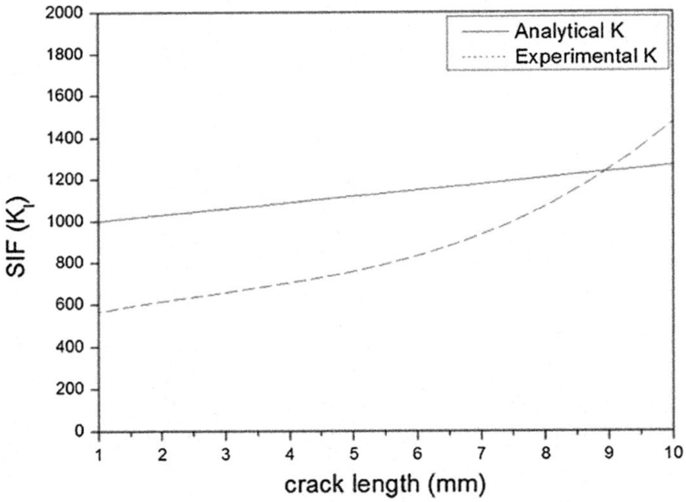

(b) SIF versus fatigue crack length of the CCT specimen

Fig. 9 Comparison of SIF between experiment and analysis

growth because the experimental SIF value is strongly affected by the machined notch condition.

Once SIF values are obtained from the experiment, the fatigue crack propagation life is predicted based on the Paris law. The fatigue crack growth data in terms of da/dN–ΔK curves are shown in Fig. 10. In the case of the SENT specimen, material constants C and m are 2.23E-17 and 4.7, respectively. Material constants C and m of the CCT specimen are 1.90E-8 and 1.34, respectively. It should be noted that the SIF values are directly measured from the experiment using the MFC sensors. Based on LEFM, the fatigue crack propagation life is calculated using the material parameters C and m. The predicted fatigue life is compared to the actual life of the test specimens as shown in Fig. 11. In the case of the SENT specimen, the fatigue crack propagation life obtained by experimental data is 269,498 cycles, and the calculated life is 219,146 cycles. In case of the CCT specimen, the fatigue crack propagation life obtained by experimental data is 69,083 cycles, and the calculated

Fig. 10 da/dN–ΔK curve for total test data

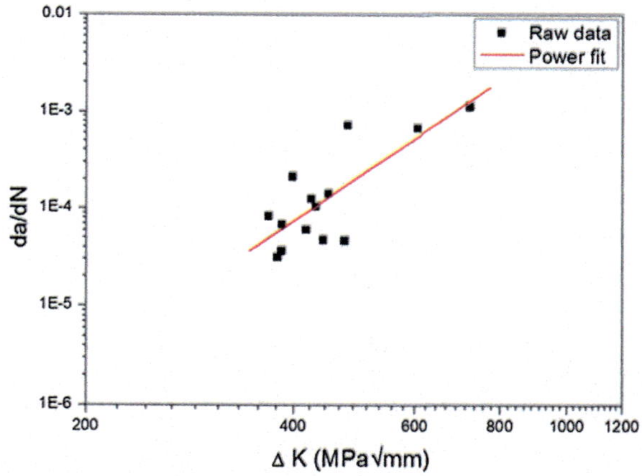

(a) da/dN - ΔK curve for the SENT specimen

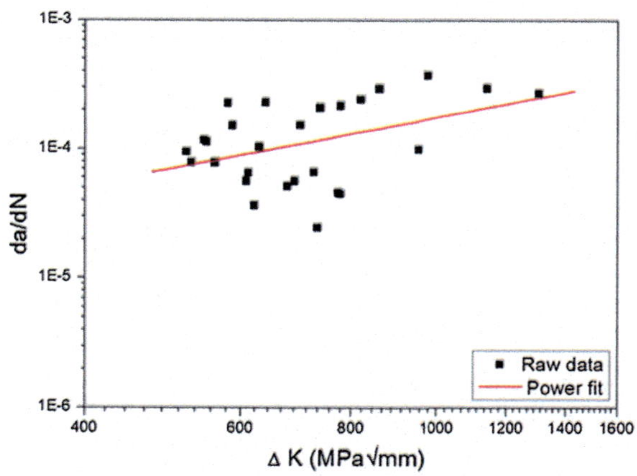

(b) da/dN - ΔK curve for the CCT specimen

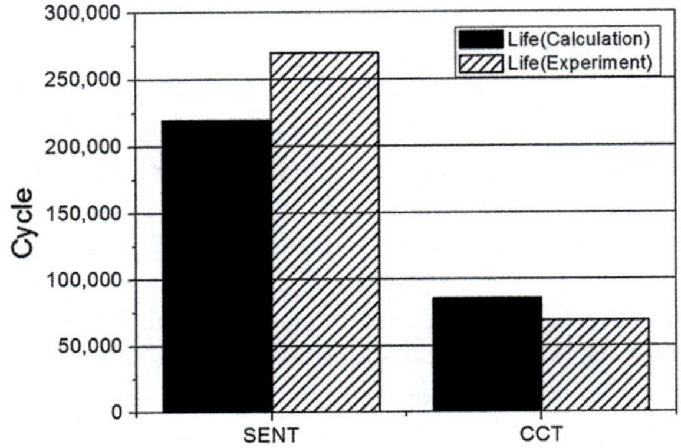

Fig. 11 Comparison of crack propagation life between experiment and prediction

life is 85,786 cycles. The estimation error is determined to be within 24%. Although it requires more experiments with different types of cracks, the technique appears promising for fatigue crack detection and fatigue life prediction.

6 Discussion

If a cyclic load applied to a structure containing a crack is too high, the crack may suddenly grow and cause structural failure. In order to ensure the structural integrity in the presence of a crack, a specific parameter SIF is typically used to define the stress state of the crack tip. For a given material, brittle fracture will occur when K_I reaches the critical value K_{IC}. In other words, structural failure associated with fracture can be estimated by continuously monitoring SIF values in real time.

As mentioned previously, the SIF is an important parameter for evaluating structural integrity. However, it is actually difficult to detect the fatigue crack growth and to measure the K_I in real time. This study demonstrated that SIF can be measured in real time using a set of MFC sensors. Based on the approach from this study, a conceptual schematic for the failure assessment is illustrated in Fig. 12.

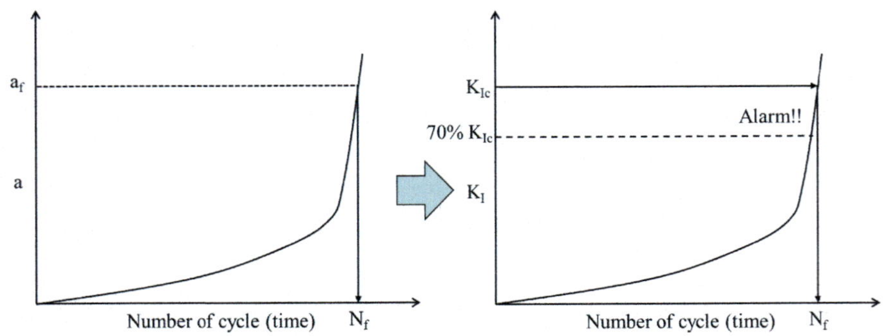

Fig. 12 An example of the failure assessment

We can predict structural failure when fatigue crack and K_I reach the critical value, such as the final failure crack length a_f and the fracture toughness K_{IC}. In this regard, if K_I can be measured using MFC sensors, we can evaluate the safety and residual life of the structure.

7 Conclusion

In this paper, an experimental study was carried out to directly measure the stress intensity factors of SENT and CCT specimens using MFC sensors. A set of MFC sensors is used to measure SIF in real time. The measured values of the SIF are later used for predicting the fatigue crack propagation life. Considering the effect of area along the direction of fatigue crack growth, the SIF method is also improved to take into account the changing stressed area of the piezoelectric effect. In addition, SIFs obtained from the experiments are compared with the results obtained from an analytical approach based on LEFM. A few important observations from this study can be summarized as follows:

- This study proposes an experimental approach that can directly measure the SIF as a fatigue crack grows. One of the piezoelectric materials called an MFC sensor is employed for the fatigue crack growth test. It is possible to detect fatigue crack propagation under cyclic loads by measuring the output voltages from the sensors. This study demonstrated that the output voltages gradually change as the fatigue crack grows and eventually rapidly drop near the final fracture.
- The SIF is an important parameter that is used to evaluate the safety or failure of a structure. In particular, it is essential for a precise prediction of fatigue crack propagation life. Using the measured SIF obtained by a pair of MFC sensors, the fatigue crack propagation life is calculated and it is later compared with the experimental result. It has been demonstrated that the measured SIF can predict the fatigue crack propagation life within the maximum error of 24%. Based on the test results, the SIF measurement approach using MFC sensors can be considered as an effective SHM system.
- In this study, it was shown that a special parameter SIF can be measured using MFC sensors in real time. Based on the LEFM, we can determine the integrity of the structure by continuously monitoring the SIF in real time. When the measured SIF values approach the critical value K_{IC}, an alarm system may trigger to notify the status of the structure.

Acknowledgements This work was supported by the National Research Foundation of Korea (NRF) grant funded by the Ministry of Science, ICT and Future Planning (MSIP) of the Korea government through GCRC-SOP (No. 2011-0030013). This work was also supported by the Human Resource Training Program for Regional Innovation and Creativity through the Ministry of Education and National Research Foundation of Korea (no. NRF-2014H1C1A1073088). The authors gratefully acknowledge this support.

References

1. Farrar CR, Worden K (2007) An introduction to structural health monitoring. Philos T R Soc A 365:303–315
2. Sun FP, Chaudhry Z, Liang C, Rogers CA (1995) Truss structure integrity identification using PZT sensor-actuator. J Intel Mat Syst Str 6:134–139
3. Chaudhry Z, Joseph T, Sun F, Rogers C (1995) Local-area health monitoring of aircraft via piezoelectric actuator/sensor patches. In: Smart Structures and Integrated Systems, SPIE Conference, Proceedings of the SPIE1 268, San Diego, CA
4. Fujimoto Y (2003) Piezoelectric sensor for stress intensity factor measurement of two dimensional cracks. Eng Fract Mech 70:1203–1218
5. Butrym B, Kim MH, Inman DJ (2010) Fatigue life estimation of structural components using MFC sensors. Strain 48:190–197
6. Sumant PS, Maiti SK (2006) Crack detection in a beam using PZT sensors. Smart Mater Struct 15:695–703
7. Ryles M, Ngau FH, Mcdonald I, Staszewski WJ (2008) Comparative study of nonlinear acoustic and Lamb wave techniques for fatigue crack detection in metallic structures. Fatigue Fract Eng Mater Struct 31:674–683
8. Staszewski WJ, Lee BC, Traynor R (2007) Fatigue crack detection in metallic structures with Lamb waves and 3D laser vibrometry. Meas Sci Technol 18:727–739
9. Zagrai AN, Giurgiutiu V (2001) Electro-mechanical impedance method for crack detection in thin plates. J Intell Mater Syst Struct 12:709–718
10. Zhang SZ, Yan YJ, Wu ZY (2007) Electric potential detection for structural surface crack using coating sensors. Sens Actuators A 137:223–229
11. Paris PC, Erdogan F (1963) A critical analysis of crack propagation laws. J Basic Eng 85:528–533
12. Ashby MF (2010) Materials selection in mechanical design. Butterworth-Heinemann, UK
13. Tada H, Paris PC, Irwin GR (1985) The stress analysis of cracks handbook, 2nd edn. Pergamon Press, Japan
14. Sodano HA, Park GH, Inman DJ (2004) An investigation into the performance of macro fiber composites for sensing and structural vibration applications. Mech Syst Signal PR 18:683–697

MMM Applied for Online Monitoring of Fatigue Crack Propagation in an Industrial Steel Structure

Anatoly Dubov, Alexander Dubov and Sergey Kolokolnikov

Abstract The results of an experimental investigation using the metal magnetic memory (MMM) method are presented. A fatigue crack in an industrial press was cyclically loaded, and the stress concentration zone (SCZ) was measured at the crack location as a function of the number of cycles. Data taken during the cyclic loading process is presented showing the variation of MMM parameters as a function of crack propagation. This data leads to the conclusion that the self-magnetic leakage field (SMLF) intensity and the field gradient vary in the course of accumulation of fatigue damage in the microstructure under cyclic loading. There is an abrupt variation of the SMLF intensity and the field gradient at the same time as crack propagation was detected. The results showed that MMM is a powerful method that shows a correlation between the SCZ measured on industrial critical equipment and fatigue crack propagation.

Keywords Fatigue tests · Monitoring system · Stress · Strain · Stress distribution SHM · Structural health monitoring

1 Introduction

At present, much of the industrial equipment in Russia is outdated and requires modernization. However, the issues of outdated equipment modernization or replacement, in most cases, cannot be solved quickly for many reasons including process and economic factors. In such circumstances, nondestructive testing (NDT) challenges not only the task of detecting physical continuities, which are inadmissible under the relevant standards and regulations, but it also has to assess the possibility to operate the equipment in the presence of such defects. In this case,

---

V-1674-15.

A. Dubov · A. Dubov · S. Kolokolnikov (✉)
Energodiagnostika Co. Ltd, Moscow, Russia
e-mail: serg_kolok@mail.ru

B. Chapuis and E. Sjerve (eds.), *Sensors, Algorithms and Applications for Structural Health Monitoring*, IIW Collection, https://doi.org/10.1007/978-3-319-69233-3_7

the task of ensuring monitoring of industrial equipment's actual state by means of NDT becomes relevant.

Paper [1] presents the results of the study of steel specimens "strain-failure" process under the conditions of cyclic load, which experimentally confirmed the possibility to apply the metal magnetic memory (MMM) method and the tester of stress concentration (TSC) for monitoring the loaded equipment's technical condition directly in the process of operation. For instance, Fig. 1 shows the time

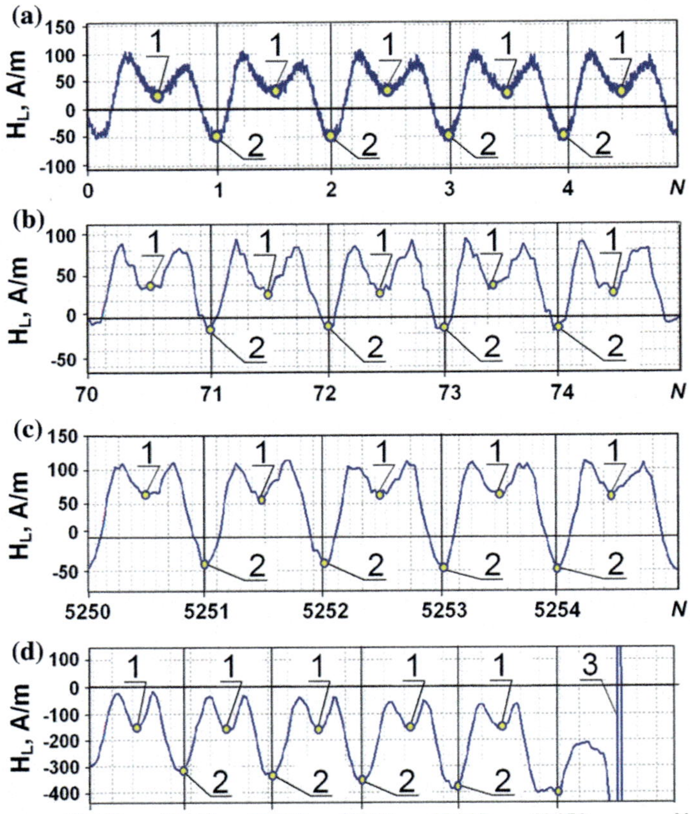

Fig. 1 Time dependence of the tangential H_L^X component of the SMLF intensity on the cyclic tensile load at the frequency of 3 Hz for the specimen made of low-carbon normal quality steel: a, b—the initial stage; c—the established mode; d—the final stage; 1—positions of the curve H_L^X, recorded at the moment of application of the maximum load P_{max}; 2—positions of the curve H_L^X, recorded at the moment of application of the minimum load P_{min}; 3—variation of the magnetic field tangential component H_L^X at the moment of the crack opening and subsequent rupture of the specimen

dependence of the tangential component H_L^X of the self-magnetic leakage field (SMLF) intensity on the cyclic tensile load, obtained during testing of the specimen made of low-carbon normal quality steel at different stages of cyclic straining right up to failure. Based on the magnetogram presented in Fig. 1, the following specific features can be pointed out:

- notable increasing of the amplitude variation of the H_L^X component of the field with increasing number of load cycles N (at the initial stage the modular variation is $|\Delta H_L^X| \approx 150$ A/m, in the established mode $|\Delta H_L^X| \approx 200$ A/m and at the final stage $|\Delta H_L^X| \approx 350$ A/m);
- the effect of abrupt decline and instantaneous growth of the field H_L^X was recorded at the maximum of the applied load P_{max} of each cycle. Finally, abrupt change of the SMLF intensity tangential H_L^X component was observed at the moment of the crack opening and subsequent rupture of the specimen.

Figure 2 presents variation of the tangential component increase H_L^X of the SMLF intensity depending on the number of load cycles N, recorded during testing of the same specimen. This curve qualitatively corresponds to fatigue curves of the metal of the specimens. Such kind of curves provides a unique opportunity to carry out the assessment of the process of fatigue time-dependent failure of the specimens depending on the number of cycles and the load frequency and amplitude.

In 2010, Energodiagnostika Co. Ltd. Experts carried out the inspection of the industrial equipment installed at the Moscow plant of “Babaev” confectionary concern. This paper considers the results of the cocoa press (shop No. 2) inspection using the MMM method and by conventional NDT methods.

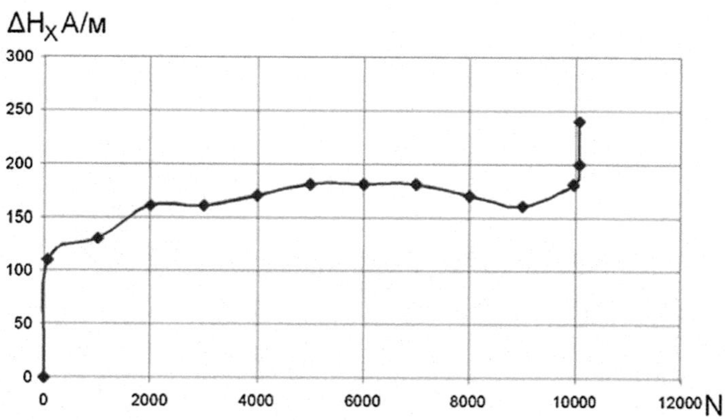

Fig. 2 Variation of the SMLF intensity tangential component increase H_L^X depending on the number of tensile load cycles [0.2; 0.95] σy

2 Inspection Using the MMM Method and Conventional NDT

The horizontal cocoa press (CP) (Type 1450/2.1) is designed to produce a solid cocoa mass. The press manufacturer is Petzholdt Heidenauer Maschinenfabrik (Germany). According to the passports data, the CP No. 2 was put into operation in 1994. Figures 3 and 4 show the general CP view. The main load-bearing (support) part of the press is the main and the auxiliary frame uprights. According to the operational service data, the press operating time is 2000 h per year in average. At the time of inspection, the CP No. 2 operating time was approximately 53,000 h with the total number of workload cycles—113,762.

The inspection program included the following stages of inspection:

- Perform 100% inspection of the press frame by the methods of visual testing (VT) and MMM to detect stress concentration zones (SCZ)—sources of damages development.
- Carrying out eddy-current testing (ET), ultrasonic testing (UT), and hardness measurement in the detected SCZs.
- Online monitoring in SCZ by the MMM method in case the developing fatigue damages are detected in order to assess the dynamics of their development.

Inspection by the MMM method was conducted on the outer surface of the main and the auxiliary upright bodies in accessible areas of the CP No. 2 in accordance with normative documents regulating the MMM method application [2–6]. The certified TSC-type instrument TSC-3M-12 with scanning devices Type-2M and Type 1-8M were used during the inspection. Figure 5 shows the inspection of the CP No. 2 main upright (Fig. 5a) by means of multiple scanning along several generating lines (Fig. 5b).

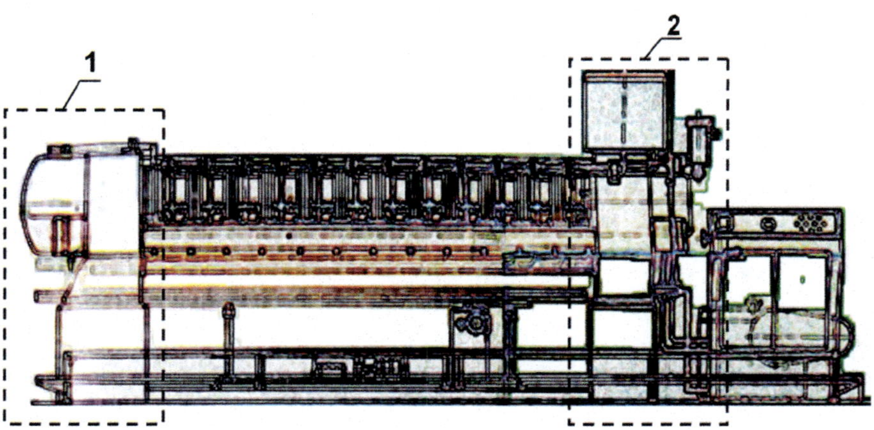

Fig. 3 General view of the coca press: 1—the main press frame upright; 2—the auxiliary press frame upright

(a) **(b)**

Fig. 4 The main (**a**) and the auxiliary (**b**) CP No. 2 frame uprights

(a) Inspection using the MMM method. *(b) Scanning pattern.*

Fig. 5 Inspection by the MMM method using the TSC-3M-12 instrument (**a**) of the CP No. 2 main upright by means of multiple scanning along several generating lines (**b**)

Inspection by the MMM method on the CP No. 2 main frame upright on the service platform side resulted in detection of four SCZs characterized by the maximum values of the magnetic field gradient dH_L/dx. The magnetograms of the SMLF intensity H_L and the field gradient dH_L/dx distributions in SCZ are presented in Fig. 6. In the detected zones SCZ-1, SCZ-2, SCZ-3, and SCZ-4, the maximum values of the field gradient dH_L/dx were $26{\cdot}10^3$, $65{\cdot}10^3$, $70{\cdot}10^3$, and $35{\cdot}10^3$ A/m^2, respectively. Figure 7 shows arrangement of the detected SCZ on the CP No. 2 main frame upright.

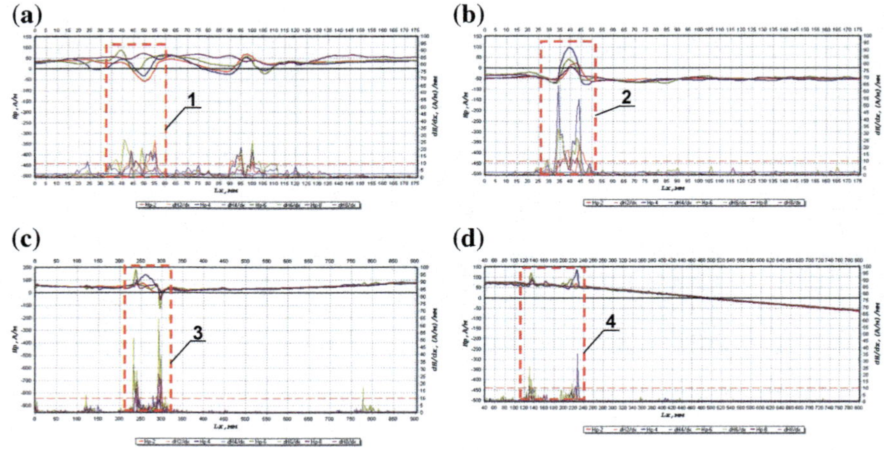

Fig. 6 Magnetograms of SMLF intensity H_L and the field gradient dH_L/dx distributions, detected in SCZ on the CP No. 2 main frame upright: 1—SCZ-1; 2—SCZ-2; 3—SCZ-3; 4—SCZ-4

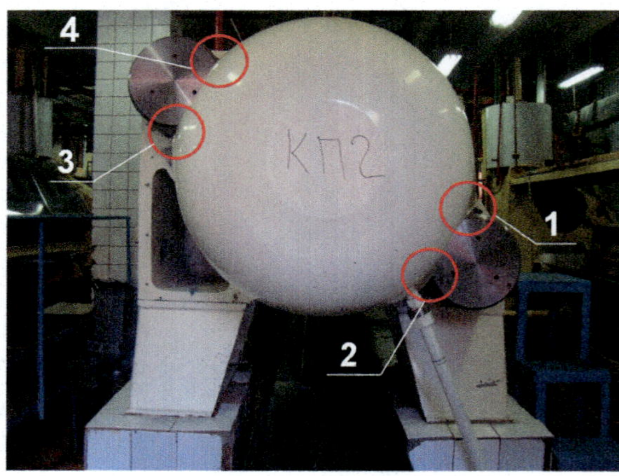

Fig. 7 Locations of SCZ detected on the CP No. 2 main frame upright on the service platform side: 1—SCZ-1; 2—SCZ-2; 3—SCZ-3; 4—SCZ-4

VT, ET, UT, and metal hardness measurements were additionally carried out in the detected SCZ.

The cracks were revealed under the paint layer using ET in the detected SCZ. After the paint removal, VE confirmed the presence of the cracks with opening width up to 0.5 mm and length of 20–100 mm.

Additional UT in the maximum crack zone in SCZ-1 detected that the maximum crack depth was 70 mm, while the upright body thickness was 130 mm. Upon performance of strength calculations and after starting the main upright replacement, the equipment owner decided to continue the CP No. 2 operation. To prevent sudden failure of the CP No. 2 main frame upright in SCZ-1 during operation, monitoring of the dynamics of stress state variation in the crack zone was ensured.

3 Monitoring of the Dynamics of Stress State Variation in SCZ Using the MMM Method

To carry out monitoring by the MMM method near the crack mouth, flux-gate transducers were installed, which ensured continuous measuring of the SMLF intensity during the CP No. 2 operation.

Figure 8 shows the arrangement diagram of flux-gate transducers for the SMLF intensity H_L measurement in SCZ-1 near the crack. The data, recorded by the flux-gate transducers using the TSC-3M-12 instrument and a special electronic unit, was further transmitted to the PC and processed using the "MMM-System" software.

Figure 9 presents the diagram of the time variation cycle of the hydraulic system oil pressure on the press piston. Points 1, 2, 3, and 4 on the diagram are the reference points of oil pressure variation. The time of one cycle of piston load is 28 min.

Figure 10 presents the magnetogram of the SMLF intensity H_L variation in the crack area for one cycle of press load. The upper part of the magnetogram shows time variations of field intensity of channels H_L-1 and H_L-2. The bottom part of the magnetogram shows the variation of gradient by base dH_L/dz that characterizes the variation rate of the field intensity between the measurement channels during the stress state variation. Reference points 1, 2, 3, and 4 on the magnetogram

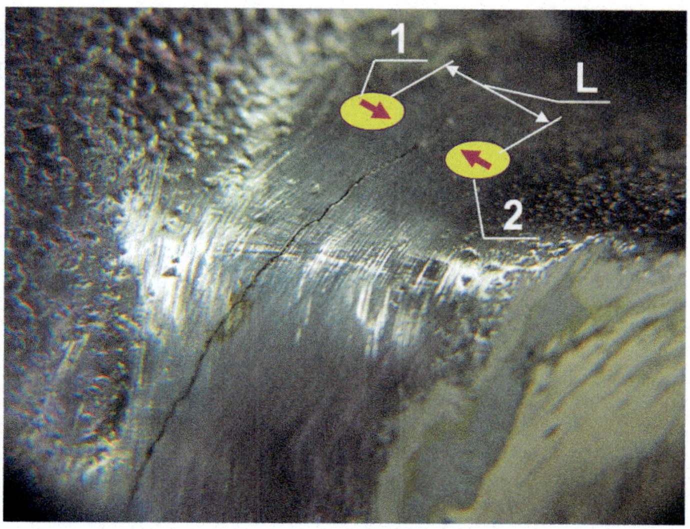

Fig. 8 Arrangement diagram of flux-gate transducers for measuring the tangential component of the SMLF intensity H_L in the crack mouth area during the press loading: 1—H_L-1 measurement channel flux-gate transducer location; 2—H_L-2 measurement channel flux-gate transducer location; L—base distance between the flux-gate transducers. Arrows show the direction of the SMLF intensity tangential component measurement

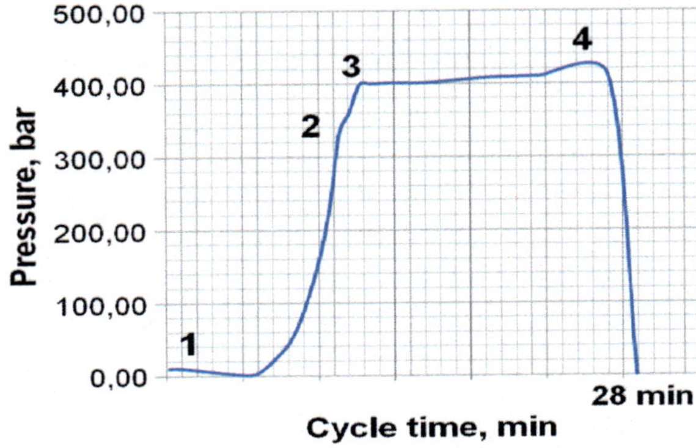

Fig. 9 Diagram of the CP No. 2 piston pressure variation cycle

correspond by time and load to points 1, 2, 3, and 4 shown in Fig. 9. Comparison of Figs. 9 and 10 shows that the graph of dH_L/dz gradient near the crack qualitatively corresponds to the variation diagram of oil pressure on the press piston.

Figure 11 presents the variation magnetograms of the gradient by base dH_L/dz between the channels H_L-1 and H_L-2 through seven cycles during the working day. The magnetogram shows that the value of the gradient dH_L/dz related to the stress state in the crack area increases with each cycle in the oil pressure growth area. This is related to the increase of the difference between the measurement channels

$\Delta H_L = |(H_L\text{-}1) - (H_L\text{-}2)|$ at the time of the frame upright's stress state variation in SCZ-1.

The results obtained during monitoring of the CP No. 2 in SCZ-1 that contains a crack can be also regarded as a crack development state in the field intensity growth area before the crack opening and specimen failure.

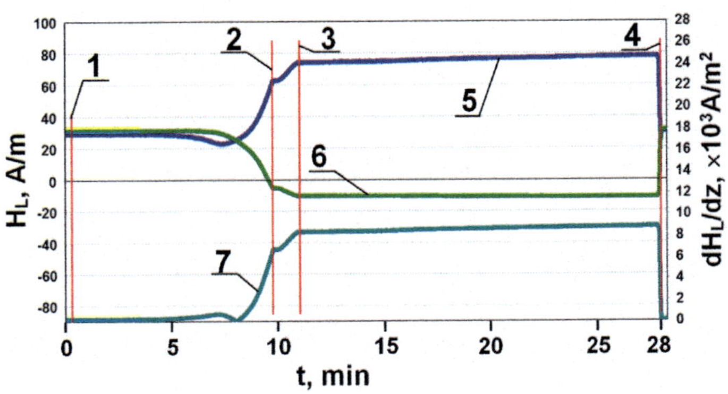

Fig. 10 Magnetogram of SMLF H intensity variation in the crack area for one cycle of press load: 1–4—reference points corresponding to the points on the pressure–time variation diagram (Fig. 4); 5—graph of H_L-1 first channel intensity time variation; 6—graph of H_L-2 s channel intensity time variation; 7—graph of time variation of gradient by base dH_p/dz between channels H_L-1 and H_L-2

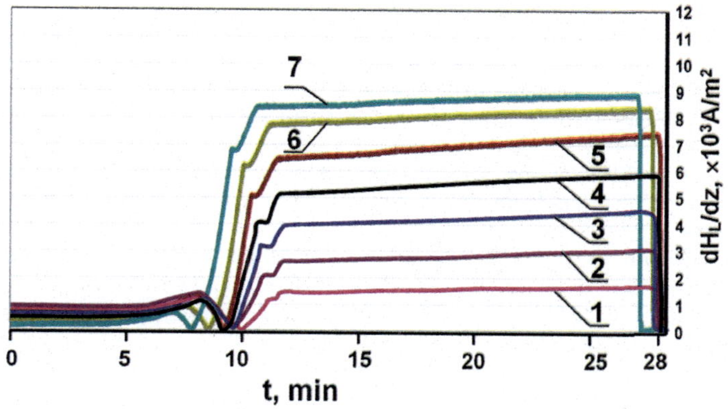

Fig. 11 Variation magnetograms of gradient by base dH_p/dz between channels H_L-1 and H_L-2, characterizing variation of the stress–strain state in the crack area in various cycles, and as a result, the continued development of cracks

As a result of these measurements, it was found that after 6–7 load cycles the SMLF intensity H_L stabilized. It was observed that each minor increase in the crack development was accompanied by abrupt variation of the SMLF intensity H_L recorded by flux-gate transducers and the TSC-type instrument.

Monitoring using the MMM method and periodic VT and UT found that at the stress state dynamic variation shown in the magnetogram of Fig. 11, the crack length and depth increase are on average 0.8 mm for every 360 press load cycles. Based on the study results, the equipment owner was recommended to schedule the press withdrawal from operation, based on the established dynamics of the stress state variation. If during the operation the dynamics of the stress state variation changes toward the sharp growth of the magnetic field intensity and hence of the gradient dH_L/dz, it was recommended to immediately stop the operation of the press for its repair.

4 Conclusions

The inspection results obtained during the press examination allow to conclude that growth of the SMLF intensity ΔH_L and the gradient dH_L/dz due to fatigue damages accumulation in the metal microstructure in the SCZ containing a developing crack.

The carried out studies confirmed the possibility to apply the MMM method for quick stress state assessment and to perform online monitoring in order to assess the fatigue damage development dynamics.

References

1. Dubov A, Kolokolnikov S (2013) The metal magnetic memory method application for online monitoring of damage development in steel pipes and welded joints specimens. Doc. IIW-2168. Weld World 57(1):123–136
2. ISO 24497-1:2007 Non-destructive testing—metal magnetic memory—Part 1: vocabulary
3. ISO 24497-1:2007 Non-destructive testing—metal magnetic memory—Part 2: general requirements
4. Russian standard GOST R 52330-2005 Non-destructive testing. Evaluation of the stressed-strained state of industrial and vehicle structures. General requirements
5. Russian standard GOST R 53006-2008 Technical diagnostics. Estimation of potential hazardous objects lifetime on the basis of express methods
6. The technique for assessment of turbine cylinders, lock and control valves condition using the MMM method. Developed by Energodiagnostika Co. Ltd.

Full-Scale Measurements and Hull Monitoring on Ships

Adrian Kahl, Hubertus von Selle and Gaute Storhaug

Abstract Full-scale measurements on ships are an important source of information for the development of ship design rules and guidelines. Several full-scale measurements have been carried out by DNV GL on container ships but also on bulk and ore carriers, LNG carriers, oil tankers, and other ships. Three measurements on container ships are presented and examples for exploitation are given. In contrast to full-scale measurements for research purpose, regular hull monitoring systems give shipboard assistance, in particular, to inform the crew about critical load levels. Next to this, hull monitoring data can also be used for onshore evaluation, e.g., of long-term hull girder loads. Hull monitoring systems are described and examples of application are given. DNV GL rules for the approval of hull monitoring systems are outlined.

Keywords Ships · Full-scale measurements · Hull monitoring
SHM · Structural health monitoring

1 Introduction

Ship design and construction are subject to change, mainly driven by economic needs, safety- and ecology-related requirements as well as technological developments. Recently, in particular, size and capacity of container ships have increased rapidly to meet the unceasing growth of marine container transport needs. Currently, container ships able to carry up to 20,000 TEU are being built. Although design experience is limited for such ships, design rules and guidelines are needed to ensure adequate structural safety and software must be developed in order to aid

V-1675-15 (XV-1494-15).

A. Kahl · H. von Selle (✉) · G. Storhaug
DNV GL, Hamburg, Germany
e-mail: hubertus.von-selle@dnvgl.com

B. Chapuis and E. Sjerve (eds.), *Sensors, Algorithms and Applications for Structural Health Monitoring*, IIW Collection, https://doi.org/10.1007/978-3-319-69233-3_8

the designer assessing the ship structure. Full-scale measurements are an important source of information for developing and validating design rules and guidelines. In particular, data is acquired that cannot be generated by computations because the respective methods or tools are not available, not mature enough, or the required effort for such computations would be excessive. An example of the latter is long-term, high-frequency loads from hull girder vibrations which are important for the ultimate and fatigue strength of the hull girder. Several full-scale measurements have been performed by DNV GL in the past on container ships, bulk and ore carriers, LNG carriers, oil tankers, and other ships. More recently, full-scale measurements have been performed on four container ships of different sizes and operated on different routes. These measurements are still running.

In contrast to full-scale measurements for research purpose, regular hull monitoring systems give shipboard assistance, in particular, to inform the crew about critical load levels. This is of benefit, especially, for large ships where the crew is often not aware of the wave and slamming loads the hull is subjected to. Hull monitoring data can also be used for evaluation of long-term loads, e.g., to assess the cumulative fatigue damage of a ship over its service life. DNV GL rules for hull monitoring set up basic requirements for the hull monitoring systems, based on which approval can be carried out and a corresponding class notation can be assigned. Through hull monitoring, currently about 100 DNV GL classed ships are monitored.

2 Full-Scale Measurements on Ships

2.1 Overview of Recent Full-Scale Measurements on Container Ships

Currently, on board of four container ships, full-scale measurement campaigns are ongoing, namely on a 4600 TEU, a 8400 TEU, a 8600 TEU, and a 14,000 TEU containership. Three of these measurements are described in more detail in the following.

Table 1 gives an overview of size, operated route, and measurement time of the three container ships. During these measurement campaigns, only the 4600 TEU Panamax (breadth $\leq$ 32.25 m) and the 8400 TEU post-Panamax (breadth > 32.25 m) container ships acquired data under conditions representing severe seaways, as their trade route spans the North Atlantic and the North Pacific. The 14,000 TEU container ship sails under conditions representing relatively mild seaways, i.e., not traveling the Atlantic and Pacific Oceans.

The measurement campaign on the 4600 TEU ship comprises the most comprehensive measurement equipment. The focus of this campaign was on the measurement of global hull girder strains, strains at hatch corners, loads on side shell longitudinals (pressures and strains), as well as global accelerations. The measurement campaign on the 14,000 TEU ship was performed within a Joint

Table 1 Overview of recent full-scale measurements

Container ship size	4600 TEU	8400 TEU	14,000 TEU
Container ship type	Panamax	Post-Panamax	Post-Panamax
Operated route	North Atlantic, North Pacific	North Atlantic, North Pacific, Europe–East Asia	Europe to East Asia
Measurement since	2008	2013	2010

Development Project with a major Korean shipyard. The sensors and cables were installed by the shipyard, and the data processing and storage units were provided by DNV GL. Here, the focus was on the measurement of global hull girder strains, strains at hatch corners, global accelerations, as well as slamming pressures. Furthermore, for this campaign, the separation of strains due to low-frequency wave loads and high-frequency hull girder vibrations was of special interest. The most recent measurement campaign for the 8400 TEU ship was triggered by the need to receive more data on the effect of high-frequency hull girder vibrations on fatigue strength for a medium-sized container ship operating worldwide. Also, for the validation of load assumption made for route-specific container stowage, data on accelerations for such a vessel is needed. Thus, the focus of this campaign was on the measurement of global hull girder strains and global accelerations.

On all ships listed in Table 1, measured data is transmitted to a central unit of the measurement system, located in the deckhouse area or the engine room. Each measurement has a different scope, and different types of sensors are installed on the ships. In Fig. 1, sensor locations and types are displayed for all three container ships.

To monitor global loads, strain gages were attached to primary structural members of the hull girders, typically at three ship stations, shown as orange rectangles in Fig. 1. The long side of the rectangle indicates the direction in which the strain is recorded. To distinguish between stress components caused by different kinds of global loads, strain gages were arranged on the 4600 and 14,000 TEU ships to enable the decomposition of stresses caused by vertical bending, horizontal bending, and torsion (see Fig. 2).

Besides global strength aspects, selected local load effects are monitored. The low torsional stiffness of container ships leads to large hatch opening deflections in oblique seas or due to roll motion, inducing high cyclic loads in the hatch corners. Thus, on the 4600 and 14,000 TEU ships, local stresses at hatch corners of the upper deck are measured by strain gages distributed along the hatch corner radius. In Fig. 1, locations at which strain gages were applied to hatch corners are indicated by green rectangles. A typical arrangement of three strain gages at hatch corners is shown in Fig. 3. As the sensors on the 8400 TEU container ship were installed on the ship in service during a voyage, no strain gages could be equipped to the hatch

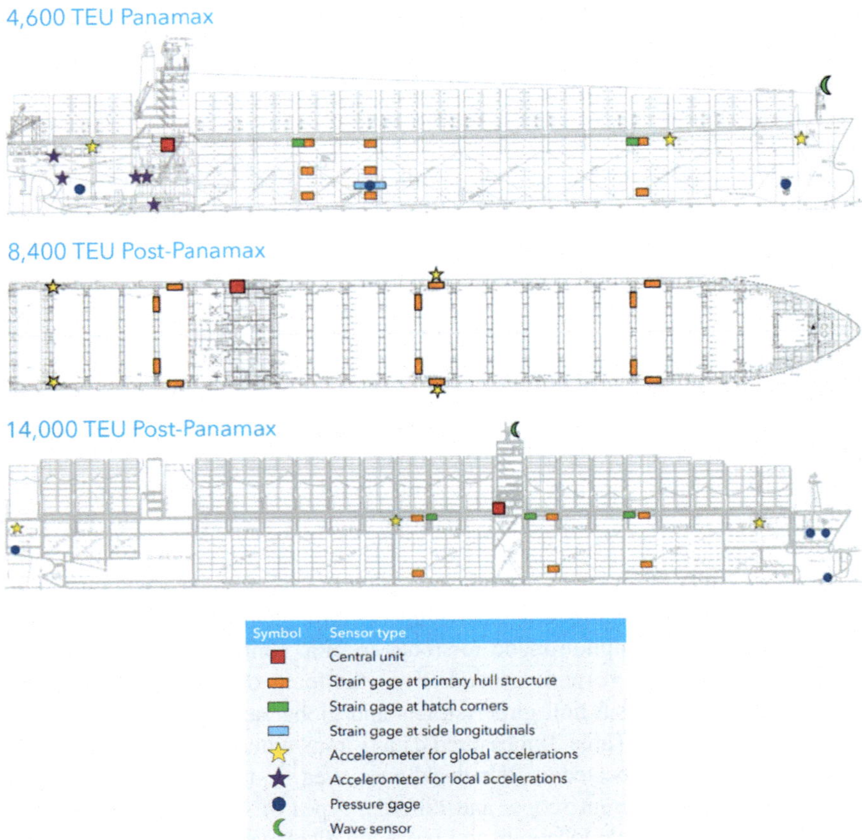

Fig. 1 Sensor locations and types

corners; instead, strain gages were attached to transverse bulkheads that are mainly loaded by torsional deflections of the hull girder.

Side shell longitudinals in way of the waterline are prone to fatigue. Reasons are fluctuating side shell pressures due to waves and roll motion of the ship as well as longitudinal hull girder stresses. On the 4600 TEU container ship, two side shell longitudinals in the midship area are monitored by strain gages. The measurement is carried out close to a ship station equipped for the measurement of global loads. The strain gages were attached at two side shell longitudinals in a wing water ballast tank, about 3–5 m below the design waterline. In Fig. 1, the approximate locations of the sensors are shown as blue rectangles. A typical arrangement of three sensors is illustrated in Fig. 4. The sensor arrangement allows decomposing stresses from lateral loads, global loads, and relative displacement of web frames.

On all ships, motions are monitored by a gyroscope located in the deckhouse area, recording vertical and horizontal accelerations, roll and pitch angles, as well as the yaw rate.

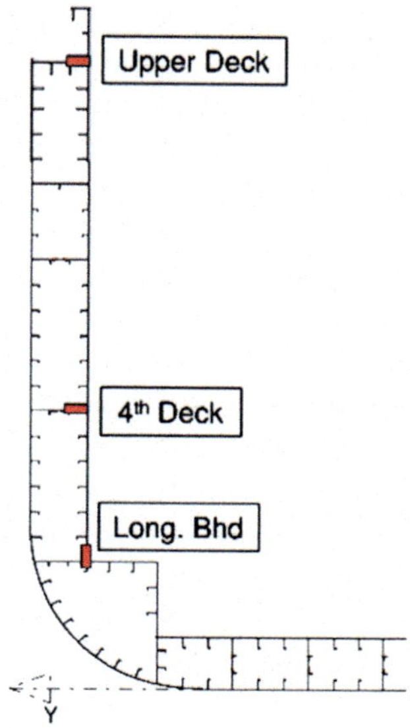

Fig. 2 Global strain measurement

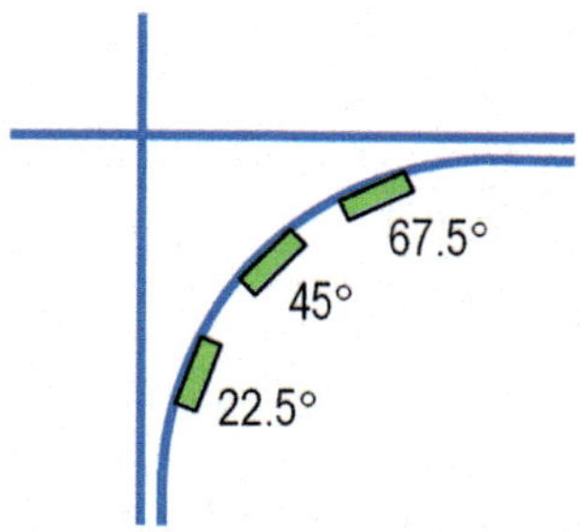

Fig. 3 Local strain measurement at hatch corners

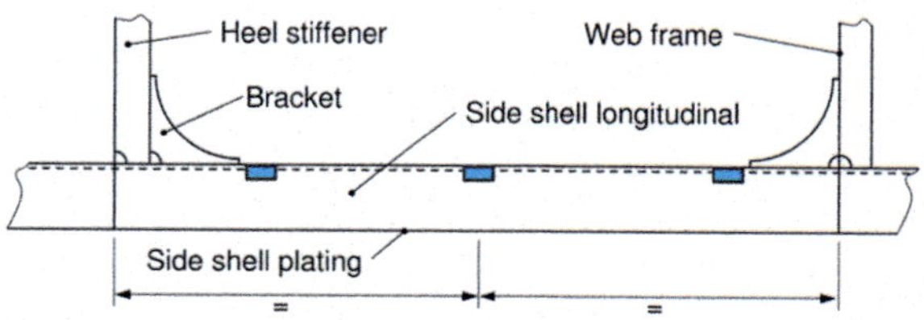

Fig. 4 Local strain measurement at shell stiffeners

To monitor accelerations at different ship stations, accelerometers were equipped as designated by yellow stars in Fig. 1. Typically, these accelerometers were installed symmetrically about a vertical plane at center line. Vertical and transverse accelerations are recorded.

Measurement of vibration of local structures is carried out on the 4600 TEU container ship at different locations in the engine room as designated by magenta-colored stars in Fig. 1. To assess the strength of the excitation source, vibration accelerations at the main engine foundation were also monitored.

On the 4600 and 14,000 TEU container ships, sea pressures at the shell are measured. On the 4600 TEU ship, the focus is on the measurement of pressures below the design waterline. In total, six pressure gages were mounted in the side shell on portside and starboard side. On the 14,000 TEU ship, the identification of slamming events is focused in the fore- and aft ship. Three pressure gages were mounted above the design waterline on portside and one at the center line close to the bottom line. The locations at which pressure gages were installed on the ships are designated by blue circles in Fig. 1.

To allow for relating measured wave-induced global and local loads as well as ship motions to environmental conditions, on the 4600 and 14,000 TEU ships, the seaway is measured by a wave sensor as part of the installed shipboard routing assistance system. For this purpose, radar scanners were mounted as shown in Fig. 1. As input for the assistance system, the ship's loading condition is traced, namely cargo distribution and masses, tank levels, drafts, and metacentric height. On the 8400 TEU ship, no wave sensor and shipboard routing assistance system was installed; however, environmental and navigational data is recorded by the ship crew on bridge registration forms during each watch. Furthermore, on all ships, navigational data is recorded from the ship systems, in particular position, speed and heading.

The sample rate for the measured data is typically 50 Hz. For all measured data, maximum, minimum, and mean values as well as the standard deviation for 15-min intervals are stored. For all sensors, time series of 30 min in length are stored, triggered by the exceedance of threshold values for selected sensors. On the 8400 TEU container ship, continuous time series are additionally recorded for some sensors. Furthermore, spectra of the measured responses are stored, yielding information on response ranges and their distribution. These spectra are obtained by applying the rain flow counting method on the continuously measured data for all sensors.

To investigate the effect of whipping and springing on ship response, data measured by strain gages in the upper deck is additionally treated by a low-pass filter. Low-frequency, wave-induced loads are separated from high-frequency hull girder vibrations. On the 14,000 TEU ship, also for the hatch corners, low-pass filtered strains are recorded. The filter frequencies were chosen with respect to the lowest estimated natural frequencies of the hull girder for the vertical bending and torsional vibration modes. On the 4600 and 8400 TEU ships, the filter frequency is about 0.4 Hz. On the 14,000 TEU ship, two low-pass filter frequencies, 0.4–0.25 Hz, were chosen to cater for different natural frequencies of the lowest vertical bending and torsional vibration modes. For the low-pass filtered data, statistical data and spectra are also stored on all ships.

2.2 Evaluation with Respect to Hull Girder Vibrations

All recent full-scale measurements include monitoring of unfiltered and low-pass filtered stresses at main hull girder structures and on one ship also at some hatch corners. As explained, this serves the evaluation of loads from high-frequency hull girder vibrations which are characterized by different modes and frequencies depending on the ship properties and the kind of excitation.

Two basic phenomena can be observed, namely whipping and springing. Whipping is induced by wave impacts on the hull (bow flare, stern, and bottom slamming), leading to transient, decaying hull girder vibrations which typically occur in moderate or harsh seaways. Springing is caused by periodic wave trains which excite resonant hull girder vibrations. Springing occurs in low-to-moderate seaways in which the ratio of wave encounter frequency and the excited hull girder vibration frequency, or its inverse, is a whole number ratio. Figure 5 depicts typical sample time series for a whipping (left) and a springing (right) event obtained from strain measurements amidships below the upper deck of the 4600 TEU ship. Shown are the unfiltered signals together with their high- and low-frequency parts. The high-frequency part of the whipping event shows the characteristic transient hull girder vibration caused by a single slamming event at the ship's bow. After its initiation, damping caused the vibration to decay. The high-frequency part of the springing event reveals the characteristic harmonic hull girder vibration. Springing is contributing to fatigue damage only, while whipping contributes to fatigue damage as well as to ultimate loads.

For the 4600 TEU container ship, in Fig. 6 (left), spectra for the unfiltered and low-pass (LP) filtered stress are plotted as a normalized stress range versus the cumulative frequency for a strain gage located below the upper deck close to amidships. These stress range spectra display the influence of high-frequency loads on long-term stresses, in that high-frequency loads increased the stress ranges as well as the number of load cycles. Stress ranges of low probability of occurrence increased more than stress ranges of higher probability of occurrence because

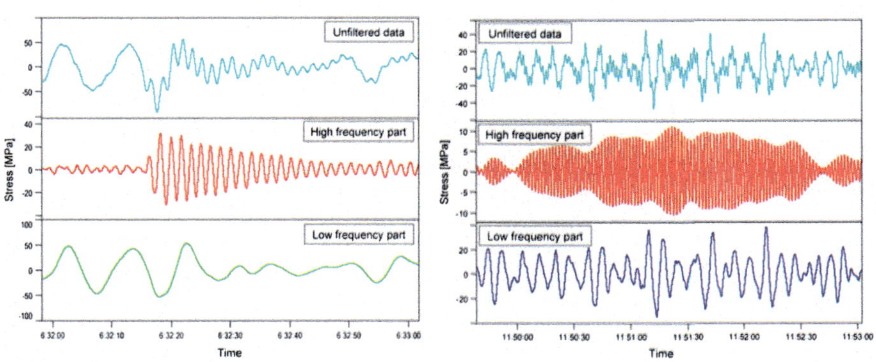

Fig. 5 Sample stress time histories containing whipping (left) springing (right)

whipping was more pronounced in the less probable, severe seaways. The increase of stress ranges of low probability of occurrence is significant for the ultimate strength (collapse strength) of the hull girder. However, due to the limited measurement time, the measured maximum stress ranges and their increase due to whipping may not be representative for the whole ship design life. The smaller stress ranges occurring between about 10^4 and 10^6 cycles, designated as "fatigue strength regime" in Fig. 6, contribute most to fatigue damage.

To quantify the effect of high-frequency loads on the life to failure of typical ship structural details, fatigue damage was determined from the measured spectra of low-pass filtered and unfiltered stresses. As a measure of fatigue damage, the cumulative damage ratio, D, was calculated according to the Palmgren-Miner linear cumulative fatigue damage hypothesis. For the stress range spectra shown in the left diagram in Fig. 6, the right diagram in Fig. 6 plots the cumulative frequency against the associated distribution of fatigue damage, here expressed as damage ratios D_i. As can be seen, load cycles in the range of 10^4–10^6 caused the major part of fatigue damage, in other words, load cycles outside this interval barely contributed to total fatigue damage.

The ratio of cumulative damage for the unfiltered and the low-pass filtered response reflects the increase of fatigue damage caused by the high-frequency part, here referred to as HF factor. As example, Table 2 lists the resulting HF factors for the 4600 TEU ship based on strains measured from gages located below the upper deck, where L designates the ship length. It shows that a large part of the total fatigue damage (up to 35%) is caused by hull girder vibrations. On larger ships, the contribution was found to be even larger with over 50%, see e.g., Storhaug and Kahl [9] and Kahl et al. [7].

Among others, based on findings as described above, DNV GL issued guidelines for "Fatigue and Ultimate Strength Assessment of Container Ships Including Whipping and Springing" [2].

Another investigation that was based on results from full-scale measurements was concerned with the question whether the fatigue resistance of welded structural details is affected by the superposition of high- and low-frequency loads and if the usual fatigue assessment would cover this. The rainflow-counting method for

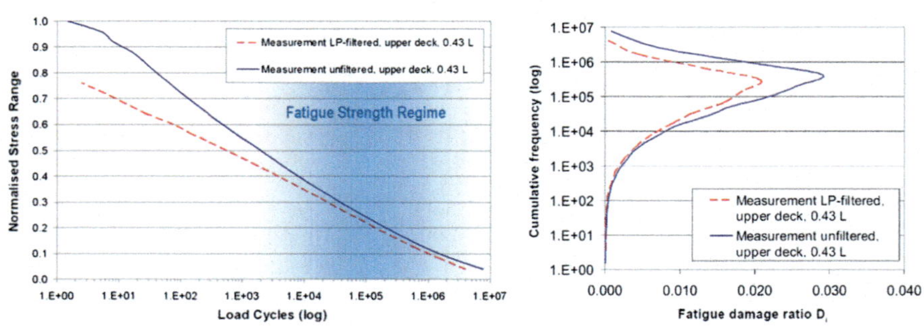

Fig. 6 Stress spectra (left) and distribution of damage ratios (right)

Table 2 HF factors for the 4600 TEU container ship

Location (L)	HF factor
0.35	1.36
0.43	1.53
0.75	1.25

classification of load cycles and the Palmgren-Miner rule for linear damage accumulation are considered to be the usual methods for fatigue assessment for loads given by stress time histories.

To investigate this topic, DNV GL and Hamburg University of Technology performed fatigue tests using load histories, measured on the 4600 TEU container ship [3, 6]. The unfiltered and low-pass filtered time series shown in Fig. 7 were repeatedly applied to welded transverse stiffener specimens. In another test program, also fatigue tests were performed for a repeated 30-min time series of which the time series shown in Fig. 7 is part of. Figure 8 shows one of the fatigue test specimens after testing.

By comparing fatigue lives from the tests to those predicted by the usual fatigue assessment methods, it was found that the Palmgren-Miner rule in combination with the rainflow-counting method is working well for both, stress histories with and without superimposed high-frequency loads. No significant positive or negative effect from the superposition of high- and low-frequency loads on the fatigue resistance of welded structural details could be identified.

2.3 *Evaluation with Respect to Container Stowage*

In 2013, DNV GL has introduced new rules that consider route-specific stowage of containers onboard container ships [5] to contribute to increasing container capacity compared to the standard lashing procedures. The standard lashing rules are based

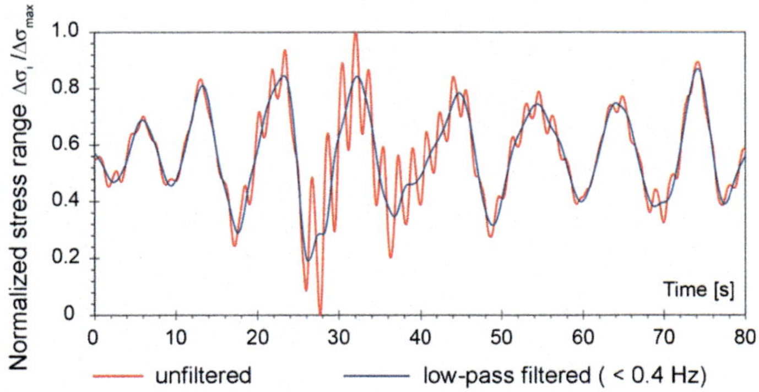

Fig. 7 Stress history applied in fatigue tests

Fig. 8 Fatigue test specimen after test

on unrestricted service in any sea environment. The new lashing rules consider route-specific lashing, allowing ships to carry more cargo, heavier containers stacked higher on deck, and to facilitate more stowage options. For the development of the new rules, systematic long-term seakeeping computations of ship motions and lateral accelerations for route-specific environments were performed [8]. These seakeeping computations in route-specific environments utilized wave statistics derived from full-scale measurements as described above.

The computed lateral accelerations for route-specific environments were validated by comparison to results from full-scale measurements on the 4600 TEU and 14,000 TEU container ships. From the measured data, Weibull spectra of transverse accelerations were extrapolated to 20 years of operation to obtain the associated extreme values for both ships (see Fig. 9 for an example).

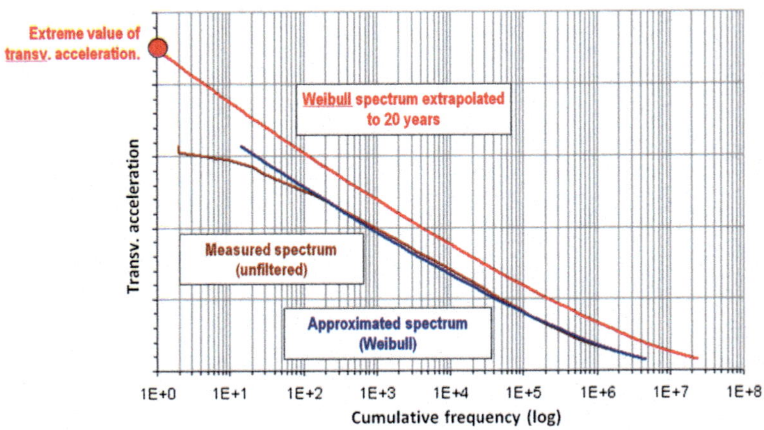

Fig. 9 Determination of extreme values of transverse accelerations

Table 3 Transverse accelerations on upper deck for two container ships

Ship (route)	4600 TEU (worldwide operation)		14,000 TEU (Europe–East Asia operation)		
Location	Aft	Fore	Aft	Amidships	Fore
Measurement	0.42 g	0.48 g	0.32 g	0.34 g	0.41 g
Rules for route-specific cargo stowage	0.45 g	0.49 g	0.40 g	0.39 g	0.42 g

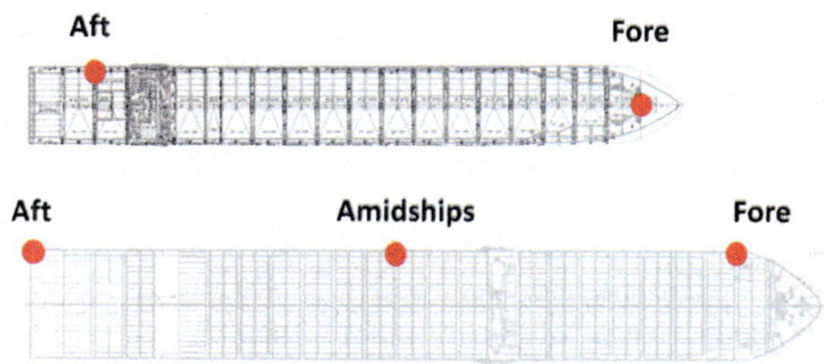

Fig. 10 Locations for the evaluation of transverse accelerations on the 4600 TEU (top) and the 14,000 TEU (bottom) container ship

Table 3 lists the resulting comparable measurement- and rule-based transverse accelerations for locations on the upper deck of the container ships as shown in Fig. 10. The generally favorable agreement validated the new rules for route-specific stowage and lashing of containers.

3 Hull Monitoring on Ships

3.1 Why Hull Monitoring?

Hull monitoring is a tool to monitor, in particular, dynamic loads on the ship hull structure. Similar to the loading computer for the static loads, the hull monitoring system can compare actual loads to threshold values that represent limitations of, e.g., structural strength. More than that, a hull monitoring system, presumed that a correct calibration has been carried out for the static condition, can also monitor the real still water loads.

Hull monitoring can close the gap between rule loads and the lack of tools to control it. Rule loads are the basis for ship hull structural design; however, good seamanship is a rule assumption. The latter implies the appropriate operation of the ship in harsh sea states (e.g., speed reduction) and that bad weather is avoided as far as possible (weather routing). Hull monitoring gives feedback to the crew on the actual hull girder loads and, thus, provides decision support to enable the crew to control them within the bounds of possibility. Table 4 gives an overview of typical load types and examples of failure modes that may be controlled by hull monitoring.

Table 4 Typical load types and failure modes

Load type	Failure mode
Hull girder wave and vibration loads	Hull girder collapse, fatigue damage
Slamming, sloshing, and green water loads	Damage to local structures (e.g., shell and tank structures)
Motions/accelerations	Cargo damage or loss

Full-scale measurements have shown that in some cases rule levels have been exceeded even in certain moderate sea states at high ship speed. The crew may not be aware of this possibility as they tend to think that wave height is the only governing parameter for hull girder loading. Also, the crew may not be aware that the permissible still water bending moment could be exceeded despite that the loading computer indicates that the still water bending moment is within the limits. A record of the half hour maxima of total hull girder stress of an LNG carrier in Fig. 11 demonstrates that the 100% warning level of the hull monitoring system has not been exceeded during more than 3 years and that the extreme values are frequently very close to the 80% warning level, suggesting that the crew actually relies on the hull monitoring system and performs seamanship accordingly.

Hull monitoring can increase safety, reduce probability of damage, and may reduce the need for maintenance. Furthermore, data can be utilized onshore to document what is happening out there, benchmark ships or operation, or establish best practice. For instance, recorded long-term data of fatigue loads can be used as a

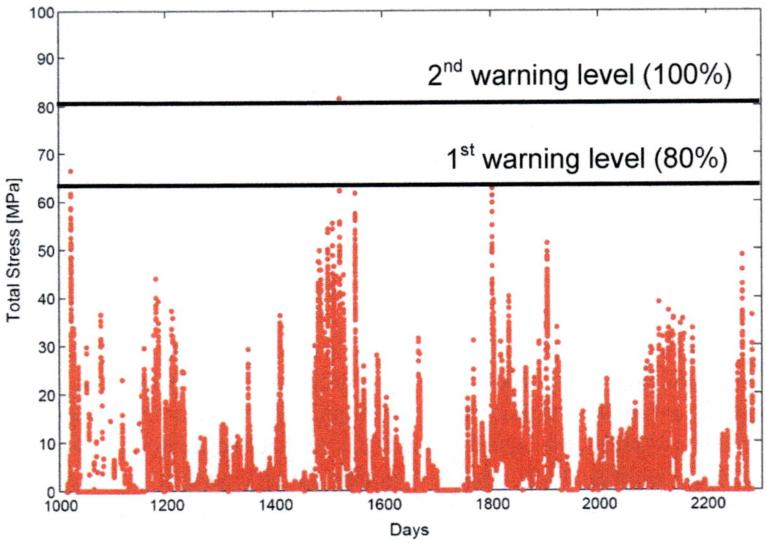

Fig. 11 Record of hull girder stress of an LNG carrier with 80 and 100% warning level of the hull monitoring system

rational basis for lifetime extension with respect to fatigue strength. Also, data may be input to decisions on future ship designs.

3.2 *What Is Hull Monitoring?*

Hull monitoring is a sensor-based monitoring system installed on board of ships. It typically comprises strain sensors and accelerometers but can include many other sensors depending on the scope of the system. Also, sensors already on board as part of other systems can be utilized for the hull monitoring. In any case, a measurement computer is part of the system with software for automatic processing, statistics, trend analysis, and warnings. Systems should also be able to store data for onshore use to enable further exploitation, if needed. A display on the bridge is the interface to the crew.

The typically required minimum scope of hull monitoring systems for displacement ships according to DNV [1] comprises monitoring of the following:

- Global longitudinal stress amidships (port and starboard side),
- Global longitudinal stress at quarter length from amidships (port and starboard side) for ships with L > 180 m,
- Loading computer data (e.g., still water bending moments),
- Position, speed, course (from ship systems), and
- Wind conditions.

The typically recommended scope comprises in addition monitoring of the following:

- Global longitudinal stress at quarter length from amidships (port and starboard side) also for ships with L < 180 m,
- Vertical accelerations at forward perpendicular at center line,
- Transverse acceleration in the 0.4 L area amidships,
- Ship motions (six degrees of freedom),
- Lateral loads at bottom near forward perpendicular (bottom slamming),
- Lateral loads at side, and
- Wave condition (e.g., wave radar).

Figure 12 gives an example of a basic sensor layout of a hull monitoring system for an oil tanker. For certain ship types, there are other minimum required and recommended sensors, e.g., for monitoring global transverse stress at transverse deck strip amidships on container ships. The recommendations for the scope of sensors are ship-specific proposals; however, there could more sensors or other sensors for any purpose and it is the owner who decides based on his specific needs. Examples are as follows:

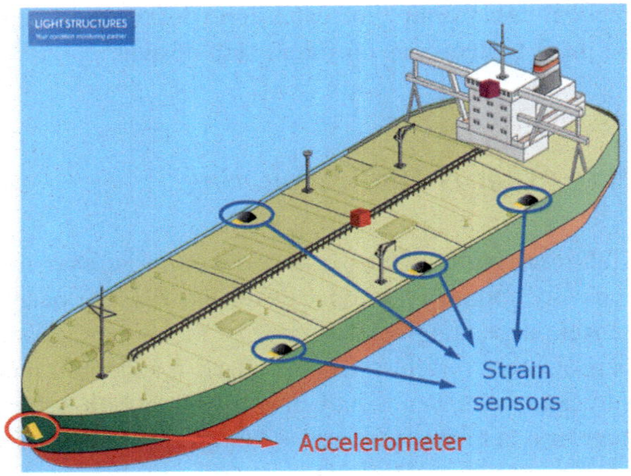

Fig. 12 Example for a basic sensor layout of a hull monitoring system for an oil tanker

- Monitoring of motion or accelerations, relevant for
  - lashing of container stacks onboard container ships,
  - sloshing in cargo tanks of tankers, or
  - comfort on board of passenger ships.
- Monitoring of shaft torque, shaft rotation, and ship speed through water with respect to fuel consumption; and
- Monitoring of ice loads on ships operating in polar regions in order to reduce the risk of damage to shell structures.

Main concerns for the minimum requirements are monitoring of still water bending and vertical wave bending as well as fatigue loads. With monitoring of dynamic hull girder loads, a hull monitoring system can be regarded as a kind of dynamic loading computer that also captures effects of hull girder vibrations (whipping and springing) that are normally not explicitly considered in design. With monitoring of still water loads, a hull monitoring system can be regarded as an alternative to the loading computer on board, which can consider additional load effects like cargo mass fluctuations. Thus, hull monitoring can verify the loading computer, which relies on calculated loads that are only as good as the input it gets, in particular, the nominal cargo weight.

A display for the hull monitoring system typically includes a main view with the most important information and several additional views that can be selected to display further detailed information (e.g., on fatigue, environment, etc.) and to operate the system. The main view typically displays still water bending moments based on measured stress and from the loading computer and puts it in relation to the limits in harbor and at sea. Also, the dynamic loading is displayed and put in relation to the design loads. Figure 13 gives an example of the main view display of a hull monitoring system for a container ship.

The measured values are compared to the given threshold values for each sensor. In the case that the value exceeds, e.g., 80% of a threshold value, an audible alarm

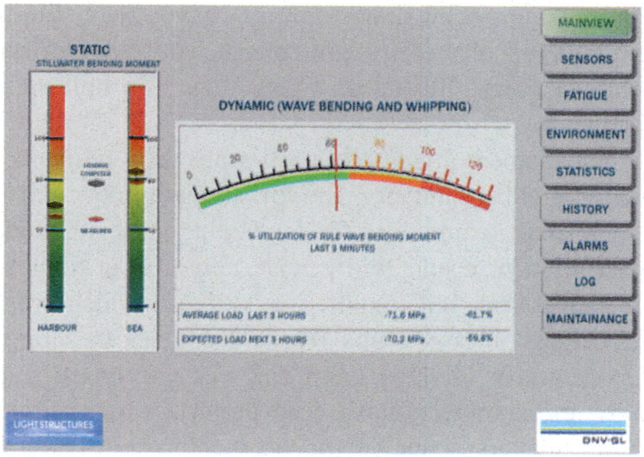

Fig. 13 Example for the main view display of a hull monitoring system for a container ship

is given and the cause of the alarm appears at the hull monitoring screen. The threshold values need to be predefined with respect to the design limits. Also, predictions of values exceeded, e.g., within the next hour, can be made based on statistics of the measured values during, e.g., the last four hours.

3.3 *Rules for Approval of Hull Monitoring Systems*

DNV GL rules for hull monitoring [1, 4] set up basic requirements for hull monitoring systems, based on which approval can be carried out. A corresponding class notation can be assigned. The rules include requirements to

- Documentation,
- Component (sensors, signal conditioning units),
- System design (primary elements, data processing, user interfaces, data storage, and extent of monitoring), and
- Installation and testing.

A required and a recommended minimum of sensors is given by DNV [1] depending on the ship type and size. The required and recommended minimum extent for displacement ships has been outlined in 3.2. However, beyond the minimum required scope, it is possible to individually configure the scope and extent of hull monitoring for each application case.

4 Conclusions and Perspective

Several full-scale measurements have been performed by DNV GL in the past, and three most recent full-scale measurement campaigns on container ships have been described. Evaluations with respect to high-frequency hull girder response and to route-specific container stowage have been illustrated as examples for the exploitation of these measurements. Hull monitoring has been explained and the benefits were illustrated. Class rules for hull monitoring systems have been outlined.

It becomes obvious that full-scale measurements are essential for backing up innovation by validation and for supplying operational data that is needed to make appropriate assumptions for developing rules and guidelines. The need for full-scale measurement results will persist; however, in future, more and more data may be available from hull monitoring systems, which are becoming more common, in particular, on large ships. This trend gives the opportunity to gather data from a large number of ships, from different ship types, and from different trades. Up to now, measurements have been performed on a limited number of ships yielding data only representative for the specific ship type, ship size, and operated route. This drawback could be compensated to some extent by the systematic acquisition and exploitation of data from hull monitoring systems. However, it has to be considered that those systems will deliver only data for the usual hull monitoring scope and cannot replace individual measurements tasks to answer specific questions.

In future, hull monitoring may be more considered also for structural design. Currently, this is already possible when hull girder vibrations are explicitly considered in ultimate strength assessment, e.g., according to DNV GL [2]. The monitored loading includes the effect of whipping, and it increases the awareness of the actual hull loading including the effect of whipping. As a consequence, the risk of overloading is reduced and the partial safety factor for wave loads including the effect of hull girder vibrations can be reduced.

References

1. DNV (2011) Rules for classification of ships, Part 6—special equipment and systems—additional class, Chapter 11—hull monitoring systems. Det Norske Veritas, Oslo
2. DNV GL (2015) Class notes no. 30.12—Fatigue and ultimate strength assessment of container ships including whipping and springing. DNV GL, Oslo
3. Fricke W, Paetzold H (2012) Experimental investigation of the effect of whipping stresses on the fatigue life of ships. In: Proceedings 11th international marine design conference. Glasgow, UK
4. GL (2015) Rules for classification and construction, I—ship technology, Part 1—seagoing ships, Chapter 5—structural rules for container ships, Annex C—certification of hull response monitoring systems for container ships. Germanischer Lloyd, Hamburg
5. GL (2013) Rules for classification and construction, I—ship technology, part 1—seagoing ships, chapter 20—stowage and lashing of containers. Germanischer Lloyd, Hamburg
6. Kahl A, Fricke W, Paetzold H, von Selle H (2014) Whipping investigations based on large-scale measurements and experimental fatigue testing. In: Proceedings 24th international symposium on ocean and polar engineering conference. Pusan, South Korea

7. Kahl A, Rathje H, Rörup J, Schellin TE (2013) Semi-empirical assessment of long-term high-frequency hull girder response of containerships—an update. In: Guedes S, Romanoff (eds) Analysis and design of marine structures. Taylor & Francis Group, London, ISBN 978-1-138-00045-2
8. Rathje H, Abt D, Wolf V, Schellin TE (2013) Route-specific container stowage. In: Proceedings 12th international symposium on practical design of ships and other floating structures. Changwon City, South Korea
9. Storhaug G, Kahl, A (2015) Full scale measurements of torsional vibrations on Post-Panamax container ships. In: Proceedings of 7th international conference on hydroelasticity in marine technology, Split, Croatia

Algorithms for an Optimal False Calls Management

Mahjoub El Mountassir, Slah Yaacoubi, Fethi Dahmene and Daniel Chauveau

Abstract This paper deals with investigations on structural health monitoring algorithms for an optimal false calls management. These false calls are caused, in the current study, by several environmental and operational factors. These factors and their effects are broken down in this paper. To demystify these effects and so reach more reliable monitoring with the best possible trade-off between probability of detection and false alarm rate, some techniques either analytical or statistical could be used. A comparative discussion between these methods is given. An example of a study using an unsupervised method is shown.

Keywords Structural health monitoring · Environmental and operational conditions · False alarm

1 Introduction

The need to ensure more human and environment safety with lower costs gives birth to Structural Health Monitoring (SHM). SHM can be defined, hence, as a discipline which aims to survey continuously the integrity of a structure. To do so, sensors/actuators should be mounted permanently to the said structure. An easy-to-use software can be used to automate the monitoring. Software decisions are based on thresholds which are closely related to the collected database. The collected data is unfortunately vulnerable to environmental and operational changes (EOC) [1], which can cause false alarms. Reducing these EOC effects is a chal-

V-1677-15

M. El Mountassir (✉) · S. Yaacoubi · F. Dahmene
Institut de Soudure, Plateforme RDI CND, Yutz, France
e-mail: m.elmountassir@isgroupe.com

D. Chauveau
Innovation Department, Institut de Soudure Industrie, Villepinte, France

B. Chapuis and E. Sjerve (eds.), *Sensors, Algorithms and Applications for Structural Health Monitoring*, IIW Collection, https://doi.org/10.1007/978-3-319-69233-3_9

lenging task. Many investigations were carried out aiming to reduce (and ideally cancel) the effects of the said EOCs [2, 3]. Different analytical or statistical algorithms are developed and applied to a database built on an experimental full-scale pipe prototype. A comparative study of the different algorithms was completed and its outputs will be concisely presented.

In this paper, first a comparative study of these different compensation methods in terms of advantages and limitations is presented. Then, an unsupervised technique called novelty detection, which has been applied to a case of study, is introduced. Concluding remarks highlighting the advantages/limitations of this technique are pointed out at the end of this paper.

2 Environmental and Operational Effects

Depending on the structure's conditions and installation environments, a variety of EOCs factors have to be considered, such as temperature and flow rate. Temperature effects have been understood as time shifts and amplitude drifts of ultrasonic guided waves signals [4]. The effects of flow rate, which is more present in tubular structures transporting fluids, were very similar and results have shown that they are as important as those caused by temperature [5].

Other factors have been investigated namely humidity, loads, and rain [6–10]. Table 1 illustrates all EOCs factors and their associated effects on guided waves which have been gathered based on a literature review performed recently by the authors.

3 Compensation and Discrimination Methods

Recently, two strategies have been developed to compensate the effects of EOCs: Optimal Baseline Selection (OBS) and Baseline Signal Stretch (BSS) [2]. As the names would suggest, both methods are optimization algorithms that rely on the minimization of the residual error between the baseline signal and the current signal

Table 1 EOCs factors and their associated effects on the collected waveforms

EOCs factors	EOCs effects on the collected waveforms			
	Time shift	Amplitude drift	Dilation	Distortion
Temperature	✓	✓	✓	✓
Flow rate	✓	✓	✗	✗
Humidity	✗	✓	✗	✗
Rain	✓	✓	✗	✗
Loads	✓	✗	✗	✓

(which may contain information related to damage). OBS utilizes multiple baseline time traces recorded for a range of conditions where the structure will operate. Each recorded signal represents a time trace measured at a specified operating condition. To find the best time trace that matches the current signal, a criterion needs to be defined. Multiple criteria can be suggested, but the most effective is the maximum residual amplitude. BSS is established on the fundamental physical understanding of the effects of temperature on guided waves. It aims to stretch the current signal to be like it was measured at the same conditions.

Another analytical method which deals with EOCs effects is the Local Temporal Coherence (LTC) [11]. It has been initially designed for overcoming temperature effects and more specifically the effect of time shift. The idea behind this method is to extract features that are sensitive to damage but insensitive to the variation of temperature. However, despite the promising results that can be obtained with these methods in laboratory controlled conditions, their real-world implementation is very restricted. This is mainly due to the fact that

- BSS and OBS must be combined with the baseline subtraction which is a very vulnerable technique.
- LTC has been only designed to compensate the effect of time shift; as a consequence, it cannot be utilized for the other EOCs effects.

An alternative solution to this problem is the use of statistical methods which seem to be more reliable and can be easily implemented. These methods can be divided into two categories: supervised and unsupervised learning:

- the former is used when different damage characteristics and realistic scenarios are available. It can be used for the identification of the type and the severity of the expected damage, but, in general, data from the damaged structure is not usually available.
- the latter does not need a priori information on the structure: long-term behavior such as, for example, its natural aging, damage either natural or accidental (shape, dimensions, orientation, type, location, etc.). Among these methods, which are not numerous, one can cite novelty detection as a promising technique.

4 Novelty Detection

Novelty detection is a statistical algorithm used generally to identify if a new measured signal is different from a set of baseline data. In the ultrasonic guided wave (UGW) technique, this algorithm is used for damage detection because the presence of damage could induce drastic changes in the received signal [12, 13]. In order to provide a solution for the problem of EOCs effects, two strategies are possible and they can be combined together:

- the baseline data should ideally be fully representative of all conditions where the structure will operate.
- features extracted from the collected signal should be sensitive (as much as possible) to damage but insensitive to the variation of EOCs [14]. It is worth noting that finding these features is challenging.

In this study, to account for some variation in these EOCs, a digital stochastic noise was added to ultrasonic measured signals. Once the data has been contaminated with noise, a threshold must be defined in order to determine whether a candidate (features from current signal) is an outlier or an inlier. Its computation depends on the baseline data distribution: parametric or nonparametric. In the case of a parametric distribution, and more particularly, if the data is assumed to represent a Gaussian distribution, it is common practice to fix the threshold value at 99.73% of its confidence limit. Therefore, if a candidate is classified as an outlier, there is only 0.27% probability of false classification. In the case of nonparametric distribution, the threshold is calculated using the extreme value theorem [15].

We have applied this algorithm to a case study where data acquisition (emission and reception of UGW) was performed using MsS System (designed for pipeline guided wave inspection). Damage was simulated by attaching magnets to the surface of the pipe, as shown in Fig. 1. This is a simplistic manner to simulate corrosion (i.e., guided waves interact with a reduction in material thickness as well as an addition in it). Magnets, which are identical to each other, are placed circumferentially to mimic circumferential corrosion. If the contact between each magnet and the pipe is the same, the size of the defect built by four magnets is four times the size of that built by one magnet. The idea behind this is to simulate progressive corrosion growth and correlate defect detectability under EOC with the defect size.

Before applying this algorithm, a user must define which type of novelty detection is needed, either univariate or multivariate analysis. The univariate analysis explores each damage-sensitive feature separately, while, in the multivariate analysis, all features must be combined in a multidimensional vector. In this case, the test of discordancy is performed using the Mahalanobis square distance

Fig. 1 Different sizes of a defect growing in the circumferential direction of the pipe

(MSD) [16] given by Eq. 1. Note that in some references, the MSD is defined without the square root operator [17]:

$$\mathrm{D} = \sqrt{(x - \bar{x})^t \cdot K^{-1} \cdot (x - \bar{x})} \qquad (1)$$

where x is the potential outlier damage index vector, $\bar{x}$ is the mean vector of the baseline, K is the covariance matrix of the baseline, and t is the matrix transpose index.

5 Univariate Analysis

Four damage-sensitive features have been extracted from the measured signals which are RMS (Root Mean Square), variance, maximum amplitude, and peak-to-peak amplitude. Results in Fig. 2 illustrate the test of discordancy in the case of using RMS. Since the baseline data is normally distributed, the threshold represents 99.73% of the Gaussian confidence limit. As it can be seen, each damage size is properly detected even in the case of one defect. The differentiation between all damage sizes is clearly noticed by the increasing steps at samples 201, 401, 601, and 801. The samples have been obtained by adding random noise to the measured

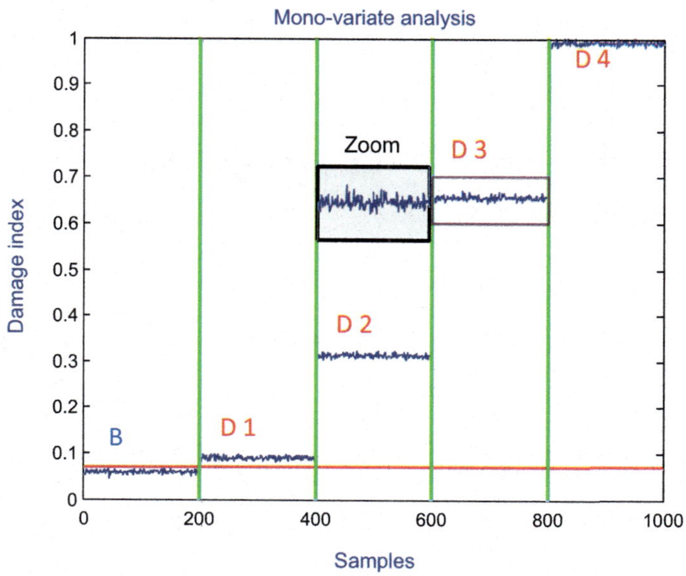

Fig. 2 Monovariate analysis based on RMS as a damage-sensitive feature. *Samples* axes indicate the number of signals. B and DI (I = 1, 2, 3, 4) correspond to Baseline and Defect I, respectively

signals. Each type of damage, including the baseline, has been represented by 200 samples. For further details, the reader is invited to refer to Ref. [18].

Despite these satisfactory results, the use of univariate analysis does not guarantee that all the damage-sensitive features will have the same response in terms of detectability. In other words, some features could for example indicate the presence of a defect and others not. This fact reduces the applicability of such a method. For this reason, a multivariate analysis is often used.

6 Multivariate Analysis

In the multivariate analysis, the selected four damage-sensitive features (this number is a choice; it can be more or less) are used simultaneously to construct a four-dimensional vector. The MSD, calculated from the baseline as well as the damaged configurations, is illustrated in Fig. 3.

For the sake of simplicity and to fairly compare the two analysis methods (mono- and multi-variate), the values of damage index were normalized. After that, the mean value of the damage index (DI) was subtracted from the threshold in order to quantify the percentage of the increase in sensitivity. This can be described mathematically using the following equation:

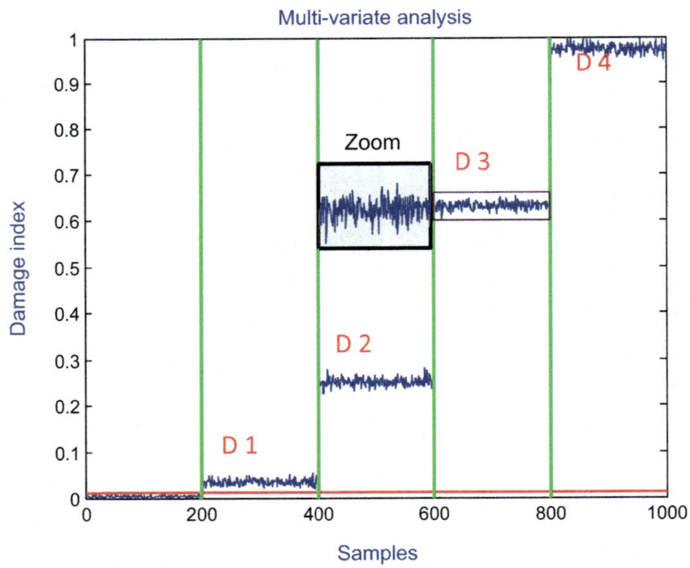

Fig. 3 Multivariate analysis obtained by extracting four features from each acquired signal

Table 2 Comparison between the monovariate and the multivariate analysis in terms of sensitivity

	Defect 1	Defect 2	Defect 3	Defect 4
Mono variate (Mean(DI)-Threshold)	0.0205	0.2416	0.5873	0.9165
Multi variate (Mean(DI)-Threshold)	0.0239	0.2422	0.6225	0.9630
PER (%)	16.58	0.24	5.99	5.07

$$\mathrm{PER} = \frac{[\mathrm{Mean}(\mathrm{DI}_{\mathrm{multi}}) - \mathrm{Th}_{\mathrm{multi}}] - [\mathrm{Mean}(\mathrm{DI}_{\mathrm{mono}}) - \mathrm{Th}_{\mathrm{mono}}]}{[\mathrm{Mean}(\mathrm{DI}_{\mathrm{mono}}) - \mathrm{Th}_{\mathrm{mono}}]} \times 100 \qquad (2)$$

where DI is the damage index value, Th is the threshold, and PER is the percentage of increase in sensitivity.

Results illustrated in Table 2 show that the sensitivity in the case of multivariate analysis has been increased by 16.58% for the first type of defect. Besides, it increases also for the other types of defects (two, three, and four defects) which confirm that the multivariate analysis performs better than the univariate analysis.

7 Conclusion

In this paper, we have reported that compensation methods which are based on the baseline subtraction are not suitable for monitoring structures operating in harsh conditions. Statistical methods may be a reliable alternative. We have investigated the use of novelty detection as a damage detection algorithm under EOCs. Results were satisfying since this algorithm was capable of detecting all types of defects. However, the following points must be considered when implementing it:

- Baseline data form (parametric or nonparametric distribution).
- Number and type of features.
- Level and kind of noise only in the case of simulation. This point is not included in real-world conditions.

References

1. Sohn H (2007) Effects of environmental and operational variability on structural health monitoring. Philos Transac Ser A Math Phys Eng Sci 365:539–560
2. Croxford AJ, Wilcox PD, Drinkwater BW, Konstantinidis G (2007) Strategies for guided-wave structural health monitoring. Proc R Soc A Math Phys Eng Sci 463:2961–2981
3. Croxford AJ, Moll J, Wilcox PD, Michaels JE (2010) Efficient temperature compensation strategies for guided wave structural health monitoring. Ultrasonics 50:517–528

4. Putkis O, Dalton RP, Croxford AJ (2015) The influence of temperature variations on ultrasonic guided waves in anisotropic CFRP plates. Ultrasonics 60:109–116
5. Eybpoosh M, Berges M, Nah YN (2014) Investigation on the effects of environmental and operational conditions (EOC) on diffuse-field ultrasonic guided-waves in pipes. In: Proceedings of international conference on computing in civil and building engineering, pp 1198–1205. Orlando, Florida, United States
6. Chen F, Wilcox PD (2007) The effect of load on guided wave propagation. Ultrasonics 47:111–122
7. Michaels JE, Lee SJ, Chen X, Shi F, Michaels TE (2011) Understanding and exploiting applied loads for guided wave structural health monitoring. In: Aircraft airworthiness & sustainment conference. San Diego, CA, 18–21 April 2011
8. Schubert KJ, Brauner C, Herrmann AS (2013) Non-damage-related influences on Lamb wave-based structural health monitoring of carbon fiber-reinforced plastic structures. Structural Health Monitoring 13:158–176
9. Schubert KJ, Herrmann AS (2013) A Compensation Method for Environmental Influences on passive lamb wave based Impact evaluation for CFRP. Key Eng Mater 569:1265–1272
10. Cicero T, Cawley P, Lowe MJS, Simonetti F (2009) Effects of liquid loading and change of properties of adhesive joints on subtraction techniques for structural health monitoring. In: Proceedings of the Review of Progress in QNDE, American Institute of Physics, vol. 28, pp 1006–1013. New York
11. Lu Y, Michaels JE (2008) Discriminating damage from surface wetting via feature analysis for ultrasonic structural health monitoring systems. Rev Prog QNDE 27:1420–1427
12. Rizzo P, Sorrivi E, Lanza di Scalea F, Viola E (2007) Wavelet-based outlier analysis for guided wave structural monitoring: application to multi-wire strands. J Sound Vibr 307: 52–68
13. Rizzo P, Cammarata M, Bartoli I, di Scalea FL, Salamone S, Coccia S, et al (2010) Ultrasonic guided waves-based monitoring of rail head: laboratory and field tests. Adv Civil Eng
14. Worden K, Farrar CR, Manson G, Park G (2007) The fundamental axioms of structural health monitoring. Proc R Soc London A Math Phys Eng Sci 463:1639–1664
15. Clifton DA, Hugueny S, Tarassenko L (2011) Novelty detection with multivariate extreme value statistics. J Signal Process Syst 65:371–389
16. Xiang S, Nie F, Zhang C (2008) Learning a Mahalanobis distance metric for data clustering and classification. Pattern Recogn 41:3600–3612
17. Barnett V, Lewis T (1994) Outliers in statistical data, 3rd edn. Wiley, New York, NY
18. El Mountassir M, Yaacoubi S, and Dahmene F (2015) Detection of structural damage using an unsupervised learning algorithm under variational environmental and operational conditions. In: 11th International conference on damage assessment of structures, DAMAS 2015. Ghent University, Belgium, August 24–26 2015

Printed by Printforce, the Netherlands